XSN2J-1

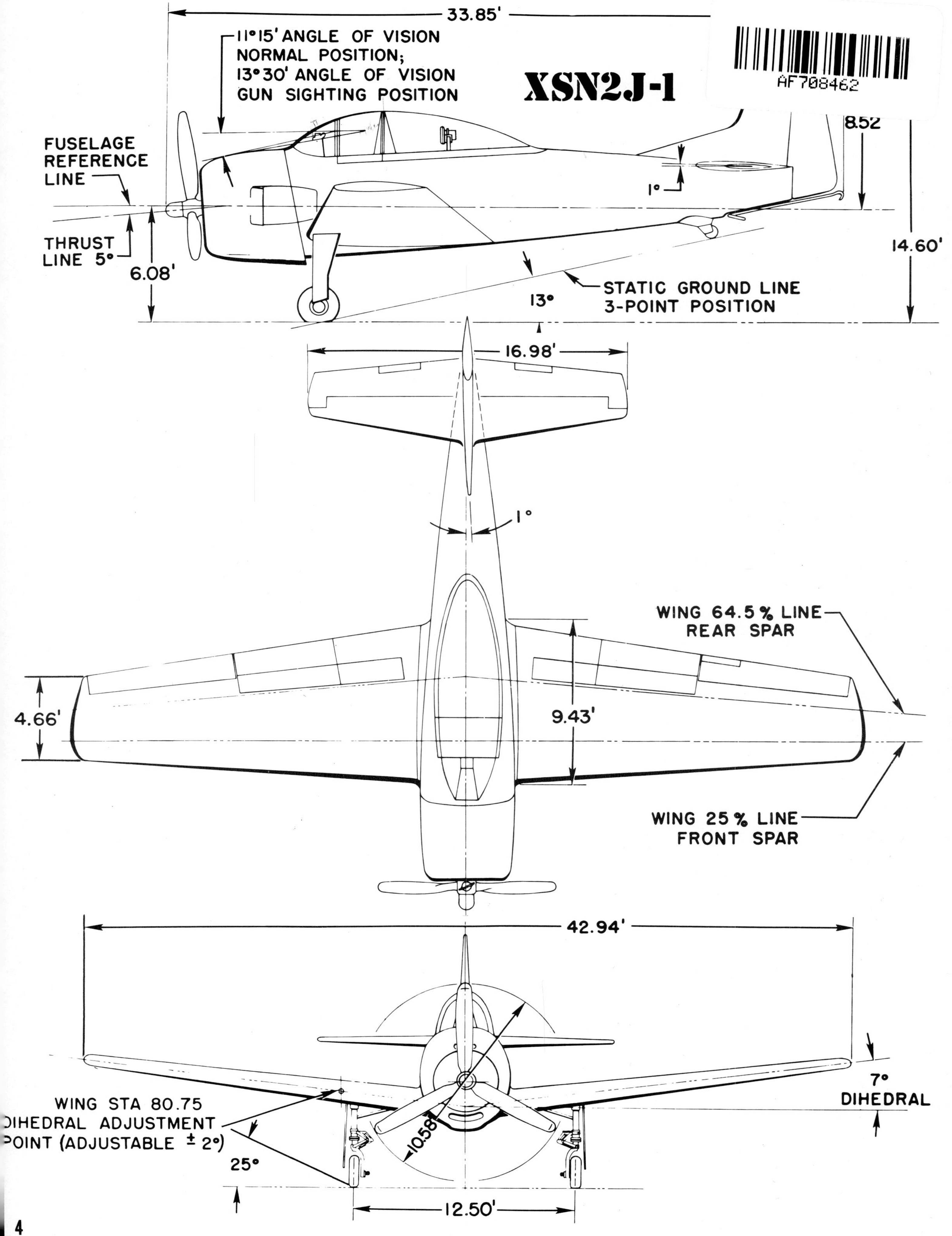

33.85'
11°15' ANGLE OF VISION NORMAL POSITION;
13°30' ANGLE OF VISION GUN SIGHTING POSITION
FUSELAGE REFERENCE LINE
THRUST LINE 5°
6.08'
8.52
1°
14.60'
STATIC GROUND LINE 3-POINT POSITION
13°
16.98'
1°
WING 64.5% LINE REAR SPAR
4.66'
9.43'
WING 25% LINE FRONT SPAR
42.94'
7° DIHEDRAL
WING STA 80.75 DIHEDRAL ADJUSTMENT POINT (ADJUSTABLE ± 2°)
10.58'
25°
12.50'

PHOTOS: Top — T-28A, 49-1571, in flight. This early T-28A does not have the traditionally painted exhaust stain area. Note wing-walk pattern and anti-glare pattern. Also note radio DF loop behind second pilot. Bottom — T-28A, 51-3763, with light grey exhaust stain area and beacon type radio antennae instead of loop. Photos via North American. (Compare these photos with those of the XSN2J-1 on page 1).

T-28A, 0-13618, with grey undersurfaces and natural metal uppersurfaces. Nose cowling, fin stripe and wing stripe day-glo orange.

INTRODUCTION

Although the T-28 is not a fighter, it has cut the teeth of many a fighter pilot and has seen its share of fighting in the ground support role. It should be noted that this will not be the last non-fighter aircraft covered under the Naval Fighters series.

Unless indicated otherwise, all drawings except the profiles are courtesy of North American. Photos used in this publication came from a variety of sources which are credited in their captions. Many photos have not been used in publications before. Anyone having photos or other information on this aircraft or other 50's era naval aircraft, may submit them for possible inclusion in future issues. Any material submitted will become the property of NAVAL FIGHTERS unless prior arrangement is made. Individuals are responsible for security clearance of any material before submission.

In this issue we are introducing a new service, photos and slides. Photos with a credit line reading, photo by Fred Roos N.F. #R____ are black and white photos. Photos with credit line reading author or N.F. collection are slides. There are close to 1,200 b & w negatives available of Naval and Coast Guard aircraft since 1969. The slides number around 5,000 from all services including Warbirds. At present, a catalog is not available, so please send all inquiries with a stamped, self-addressed envelope. Write to: S. Ginter, 1754 Warfield Cir., Simi Valley, California 93063.

ACKNOWLEDGMENTS

Ron Jawolski, Roger Besecker, Fred Roos, David Menard, Jim Burridge, Lionel Paul, North American, Robert Lawson and Don Fetters.

PHOTOS AT LEFT: Top — XSN2J-1 in flight. Note the similarities between this photo and the one on page 5. From these two photos you can see the possibility of converting the Heller or the Monogram T-28 model kits into the tail dragging XSN2J-1. Bottom — On this head-on view note the engine cowling details and the three blade propeller. Also note the main gear tires which are similar to those used on P-47s. Photos by North American.

FRONT COVER PHOTOS: Top — T-28B from NAS North Island, photographed over San Diego by Robert L. Lawson, via American Aviation Historical Society. Bottom — T-28A, 52-1205, photographed by Col. R. E. Clark via Dave Menard.

PHOTO BELOW: XSN2J-1, 121450, undergoing armament testing at Naval Air Test Center. Note 450 NATC in black under wing and rocket stubs and bomb shackles. Photo via the R.F. Besecker collection.

THE SIRE OF THE T-28

Although the U.S. Navy's T-28B and T-28C Trojan were developed from the Air Force T-28A, it's true ancestry can be traced to North American's XSN2J-1 of 1948. The XSN2J-1 was an experimental intermediate trainer designed for the Navy a a carrier capable aircraft. It was also touted as suitable for taking on the duties of teaching fighter tactics, dive-bombing, instrument and navigation training. Two flying prototypes and one static test article were built and tested.

If you look at the flying photo at left of the XSN2J-1 you can see the family resemblance. Basically the XSN2J-1 differs from the T-28 in being a tail dragger.

XSN2J-1 SPECIFICATIONS

TYPE: Two-seat Experimental Intermediate Trainer.

WINGS: Low-wing cantilever monoplane. All-metal two-spar stressed-skin structure. Flaps on underhung hinges between ailerons and fuselage. Dive brakes on upper surfaces. Dihedral angle from roots 7°, but from points outboard of landing gear attachments angle may be adjusted + or – 2°. Max. chord 9 ft. 5 in., minimum (tip) chord 4 ft. 8 in. Gross wing area 302.64 sq. ft.

FUSELAGE: All-metal stressed-skin structure.

TAIL UNIT: All-metal stressed-skin structure. Statically and dynamically balanced movable surfaces. Controllable trim-tabs in elevators and rudder. Tailplane incidence 1°. Span of tail 16 ft. 11 in.

LANDING GEAR: Retractable main wheels, fixed tail-wheel. Hydraulic retraction. Bendix air-oil shock struts. Bendix wheels and rotor-disc hydraulic brakes. Non-retracting full-swivelling tail-wheel. Static three-point ground angle 13°. Main wheel track 12 ft. 6 in. Deck-arrester hook aft of tail-wheel.

POWER PLANT: One 1,100 h.p. Wright Cyclone R-1830-78 nine-cylinder radial air-cooled engine with single-stage two-speed supercharger. Hamilton Standard Hydromatic three-blade airscrew 10 ft. 9½ in. diameter. Main fuel in wings. Total internal capacity 260 U.S. gallons. Two external drop tanks may be carried, each tank holding 58 U.S. gallons.

ACCOMMODATION: Tandem cockpits under power-operated two-piece "bubble" canopy which slides aft to provide openings above each cockpit. Pupil in front. Both cockpits similar and conform to standardised "functional" cockpit developed jointly by U.S. Army and Navy and R.A.F. Instructor can "upset" certain controls and instruments in pupil's cockpit to simulate possible emergencies. Three switches spill gyro instruments and fuel quantity gauge and five push-buttons upset airspeed indicator, altimeter, turn-and-bank, and oil and fuel pressure gauges. When switches and buttons released, instruments return to normal. Emergency system for opening canopy and releasing shoulder straps.

ARMAMENT: Two wing-mounted .50-in. machine-guns; camera gun; bomb and rocket carrying equipment.

DIMENSIONS:
Span 42 ft. 11¼ in.
Length 33 ft. 10 in.
Height (over tail — flying position) 14 ft. 7 in.

WEIGHTS AND LOADINGS:
Weight empty 5,948 lb.
Weight loaded 8,406 lb.
Wing loading 27.8 lb./sq. ft.
Power loading 7.65 lb./h.p.

PERFORMANCE:
Max. speed 308 m.p.h. at 21,750 ft.
Stalling speed 72 m.p.h.
Initial rate of climb 2,000 ft./min.
Service ceiling 32,200 ft.
Normal range 1,600 miles

T-28 SOURCE MATERIAL

BOOKS

1. Berger, Carl. *The United States Air Force In Southeast Asia.* U.S. Government Printing Office.
2. Brazelton, David. Profile Number 155. *The North American T-28.* Leatherhead, Surrey, England: Profile Publications, 1967.
3. Bridgman, Leonard/Taylor, John. *Janes All The Worlds Aircraft,* various years. Sampson Low/McGraw Hill.
4. Fitzsimons, Bernard. *Weapons And Warfare, Vol. 23.* Phoebus Publishing Co.
5. Momyer, William. *Air Power In Three Wars (WWII, Korea, Vietnam).* U.S. Government Printing Office.
6. Swanborough, Gordon & Bowers, Peter. *United States Navy Aircraft Since 1911.* Funk & Wagnalls.

MAGAZINES

1. *Air Classics.* August 1972, "Chitty-Chitty-Bang Bang", O'Hara.
2. *Air Combat.* March 1981, "Turbo Trojan", Charlie Sapp.
3. *Aero 'N' Photos.* Volume 1, number 1, May 1979, "North American T-28 Trojan", Lionel Paul.
4. *Air Fan.* Number 28, February 1981, "Trawing-5NAS Whiting Field", Jean-Michel Guhl. (Excellent French periodical.)
5. *Air Progress.* August 1971, "Inside Pensacola's Pressure Cooker", William Kemp.
6. *Approach.* August 1970, "T-28 Overstress".
7. *Flying Review International.* Volume 19, number 6, March 1964, "Air Operations Over Viet Nam, James Sunderman.

BOOK AND MAGAZINE SOURCES FOR PROFILES NOT LISTED ABOVE

1. *Air Power, The Worlds Air Forces.* Ziff-Davis Publishing, page 283, T-28S Fennec, Argentina Navy.
2. *American Aviation Historical Society.* Volume 16, number 3, Fall 1971, page 212, T-28D, 49-1591, 606th SOS. Volume 20, number 1, Spring 1975, page 72, T-28B, 13854, VF-21.
3. *The Observer's World Aircraft Directory.* Warne publishing, page 18, T-28A, 151, Cuban Air Force.
4. *The Ships And Aircraft of the U.S. Fleet.* Naval Institute Press, 7th edition, page 39, T-28B, HMS-13. 8th edition, page 37, T-28B, 130019, VT-30.
5. *U.S. Marine Corps Aircraft,* 1914-1959. William T. Lrkins, page 169, T-28B, 140008, VMIT-10.

Note on 1/72 scale drawing of XSN2J-1. After having this company drawing reduced to scale, it was noted that one measurement does not figure properly. The wingspan comes out to the same length as the T-28s when all specs. show it to be almost three feet longer.

AIR FORCE T-28A

The T-28A came to life after North American design NA-159 won the Air Force's design competition for a new basic trainer. Two XT-28s (originally XBT-28) were ordered in May 1948. On 24 Sept. 1949 the first XT-28 took to the air and shortly thereafter North American received an initial order for 266 T-28As. Eventually a total of 1194 T-28As would be built with the last one being delivered in 1953.

The first production T-28A entered service in April 1950 and became the advanced jet trainer from which students graduated to T-33s. 1956 would find the T-28A being retired and replaced by the new T-37, thus creating the Air Force's all jet training syllabus. Some T-28As lingered on into 1959 with Air National Guard units. These units had retired their F-51D Mustangs, but could not re-equip with modern jets till their runways were suitably lengthened.

The T-28A was fitted with a Wright R-1300 engine of 800 h.p.. This engine did its job, but really did not provide enough power for the big T-28. A notable design feature of the T-28A was its tricycle landing gear. Another was the nose gear, which was fitted with a hydraulically boosted, steerable nose wheel. Another important feature was the fact that provisions were made which enabled the aircraft to be configurated for armed weapons training. Thus configured, one single .50 caliber gun pack could be mounted under each wing. These pods consisted of one type M-3 .50 cal. machine gun equipped with a type J-4 gun heater and a type B-1 gun charger and airbottle. The pods would hold 100 rounds of ammo and retain the spent cases in the pod. In addition to the gun packs, S-2 bomb pylons could be fitted and used to carry either conventional bombs or the T-1 practice-bomb container or a three-tube rocket launcher.

Physically there were some differences between "A" models, as outlined below. The two prototypes carried functional speed brakes which were not present on production models. T-28s 49-1709, 51-3763 through 51-3769, and 51-7482 and all subsequent aircraft had a 4 inch lower canopy with the turnover truss (rollbar) deleted. Internally, an additional 52 gallons of fuel were carried on models 51-3463 and all future models.

T-28A SPECIFICATIONS

TYPE: Two-seat Basic Trainer.

WINGS: Low-wing cantilever monoplane. Two-spar all-metal structure. Trailing edge flaps between ailerons and fuselage. Total flap area 53.6 sq. ft. Gross wing area 268 sq. ft.

FUSELAGE: All-metal semi-monocoque structure.

TAIL UNIT: Cantilever monoplane type. All-metal structure. Area of tin and rudder 30.3 sq. ft., area of sailplane and elevators 59.7 sq. ft.

LANDING GEAR: Retractable tricycle type. Main wheels with Goodyear single-disc brakes retract inwardly into wings, nose wheel backwards into fuselage. United Aircraft Products air-oil shock struts. Nose wheel is steerable from either cockpit by electro-hydraulic boost control. Nose wheel also carries a movable taxi light. Main wheel track 12 ft. 8 in.

POWER PLANT: One Wright R-1300-1 seven-cylinder radial air-cooled engine with a military and take-off rating of 800 h.p. and a normal rating of 700 h.p. Aero Products two-blade constant-speed airscrew 10 ft. diameter. Fuel capacity 125 U.S. gallons.

ACCOMMODATION: Tandem cockpits beneath single bubble canopy. Duplicated flight controls and instruments. Amber screen blind and night flying equipment with ultra-violet and infra-red instrument lighting. Armament, gunsight, radio, etc. easily installed and removed.

DIMENSIONS:
- Span 40 ft. 1 in.
- Length 32 ft.
- Height overall 12 ft. 8 in.

WEIGHTS AND LOADINGS:
- Weight empty 5,111 lb.
- Normal loaded weight 6,365 lb.
- Max. take-off weight 6,759 lb.
- Wing loading 23.8 lb./sq. ft.
- Power loading 7.96 lb./h.p.

PERFORMANCE:
- Max. speed 288 m.p.h. at 5,900 ft.
- Cruising speed 190 m.p.h.
- Stalling speed 72 m.p.h.
- Initial rate of climb 2,570 ft./min.
- Service ceiling 29,800 ft.
- T.O. run to clear 50 ft. 436 yds.
- Max. range 1,008 miles

PHOTO BELOW: XT-28, 81371, with test probes on both wing tips. Overall color was natural metal with light grey wing tips, and rudder above top hinge. Also note small anti-glare panel and absence of aft compartment radio equipment. North American photo.

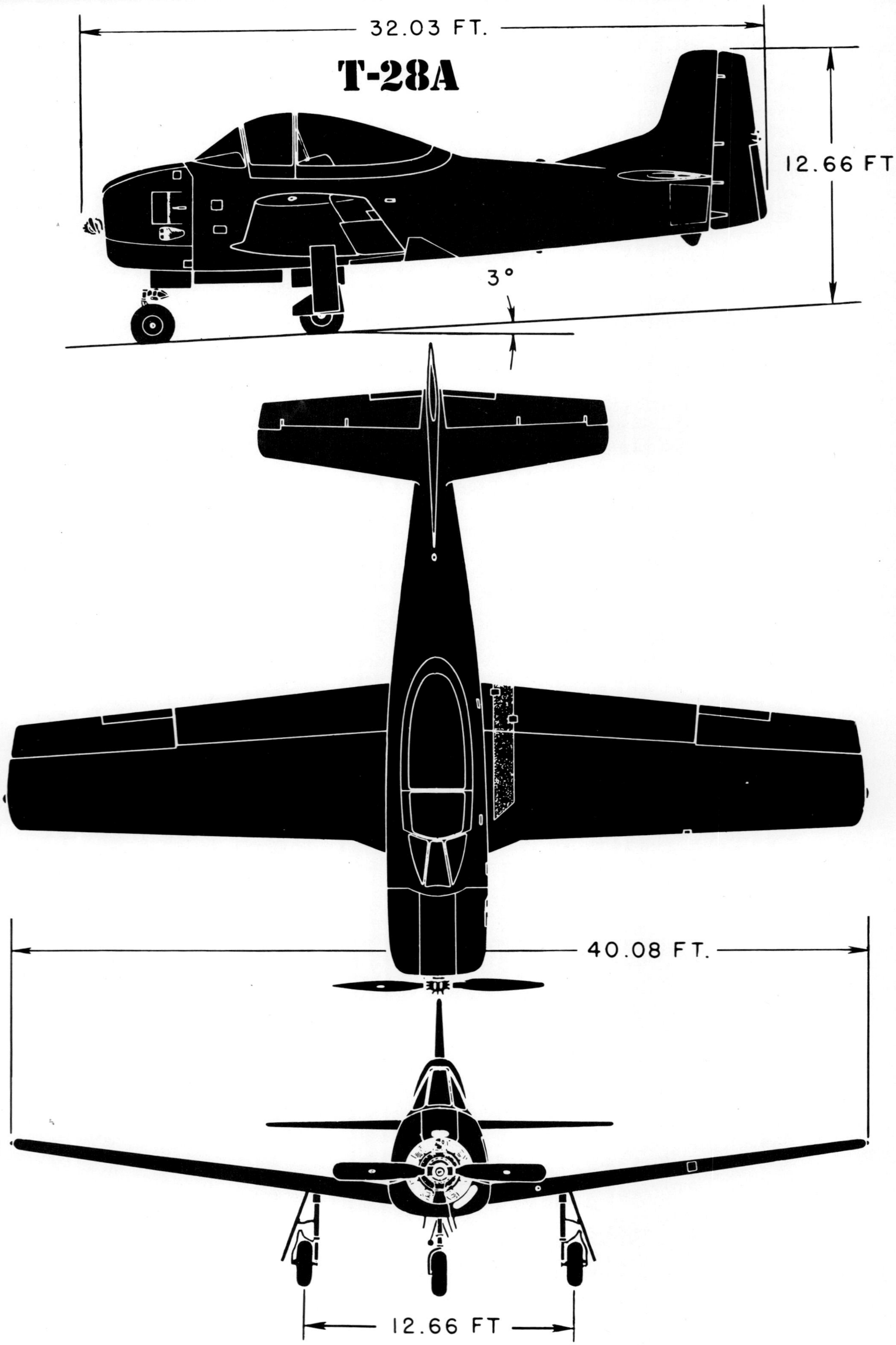
32.03 FT.
T-28A
12.66 FT.
3°
40.08 FT.
12.66 FT

PHOTOS TOP TO BOTTOM: T-28A, 50-301, at Chanute AFB in May of 1956. This is the last color scheme used on active T-28As. Light grey panel which starts aft of the stacks goes all the way to the tail. Small 301 on nose is white on a red square. Also note the separations on lettering on U.S. AIR FORCE. Photo by Dave Menard.

Two photos of T-28A, 51-3593. Note how exhaust stains have completely covered grey paint aft of wing. Note also the word BRYRN painted above insignia on tail, and that the nose wheel is black or dark grey. Photos by Balogh via Menard.

T-28A, 49-1495, at Edwards AFB on 12 July 1952. Natural metal aircraft with orange day-glo engine cowl and rudder. Exhaust stain area and top of fuselage red. Also note emblem on tail. Photo by Balogh via Menard.

T-28A, 49-1494, at Edwards AFB. Natural metal aircraft with lower half of fuselage light grey an an orange day-glo engine cowl. Blue arrow on tail with Edwards badge. AFDC badge and motto all but obliterated by exhaust stains. Note that bird has been used hard as evidenced by wrinkled skin above baggage hatch. Photo by Larkins via Menard.

T-28A, 52-205, taking off. Although slightly blurred, photo is of interest in that it shows how the gear retracks. Note main gear is almost closed but nose gear is only about half way up. Photo taken May 1956 by Col. R.E. Clark via Menard.

PHOTO BELOW: T-28A, 50-284, of the Illinois ANG. Note how light grey area extends around in front of the exhaust ports. Photo by Brazelton via Menard.

Top Photo — Beautiful in-flight close-up of T-28A. Anti-glare panel is olive green. Photo by Col. R.E. Clark via Menard. Middle Photo — One of the last T-28As in final color scheme of light grey overall with black fin tip, anti-glare panel and front end of left wing-tip. Note pilots have helmets in this photo but not in one above. Photo via Dave Menard. Bottom Photo — T-28A, NASA 502, overall light grey with black fin stripe, anti-glare panel and left wing tip. Dark grey exhaust stain area. Photo by Stephen H. Miller, via Menard 29 July 1978.

Top Photo — T-28A, 49-1492, and at least 50 more on a desert strip. Photo by Col. R. Merritt via Menard. Middle Photo — T-28A, 51-3555, in flight with CRAIG above the badge on the tail. USAF photo via Menard. Bottom Photo — T-28A, 0-17586, on 11 July 1970. All grey color scheme with black tail stripe and red fin tip. Note prop is silver on the outside and black on the inside with yellow tips. Photo by John T. Wible via AAHS.

PHOTOS AT LEFT, TOP TO BOTTOM:

Two views of T-28A, 49-1540, in early color scheme. Overall natural metal with light grey wing tips and fin tip. Prop is black on both sides with yellow tips. Also note black bottom half of landing gear door and landing light on nose gear. Photo from AAHS.

T-28A, 0-13512, in storage at MASDC in June 1972. All grey color scheme with black fin stripe and red fin tip. Photo by Roy Lock / Bruce Orriss via AAHS.

T-28A, 51-7741, overall natural metal with light grey lower fuselage and fin stripe on rudder tip. Photo by Col. R. Uppstrom via Menard.

T-28A, 52-1205, in a tight turn. Note white canvas behind aft cockpit used for instrument training and commonly called the hood. Photo by Col. R.E. Clark via Menard.

PHOTOS AT RIGHT:

T-28A, 0312, of the Ill. ANG 169th Fighter Squadron. Note 312 repeated on gear door and Indian head emblem on tail. Photo taken Jan. 1958 by Dave Menard.

Beautiful formation photo of 51-7788 and 51-7690. Photo by Brazelton via Menard.

51-7690 in flight and 51-7788 on the ground with starter cart ready. Photos by Brazelton via Menard.

INSTRUMENT FLYING HOOD

PULL HOOD FORWARD AND ATTACH STRAP TO INSTRUMENT PANEL SHROUD BELOW HANDHOLD. THE HOOD IS SPRING-LOADED TO THE STOWED POSITION.

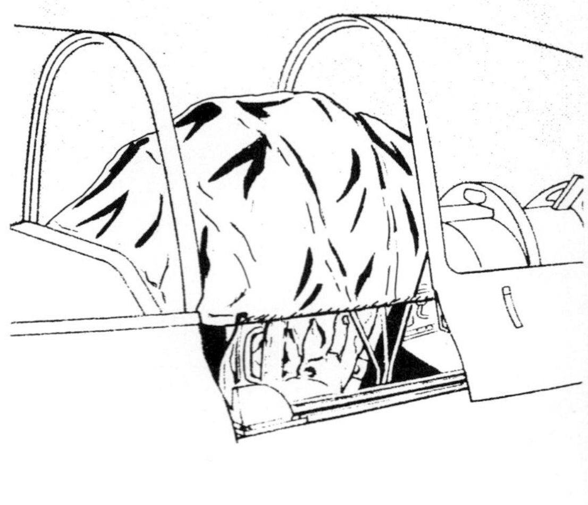

TWO PHOTOS ABOVE: T-28A, 51-7805, of the Ohio ANG. Overall natural metal color with blue tail stripe and cowling markings. Photo at right shows 17805 with military markings removed, but showing the details of the nose stripe. Photos by Baker via Menard.

T-28B In Air Force Service

After the appearance of the Navy's T-28B in 1954, a number of "B" models started showing up in Air Force markings. Other than the serial number, most T-28Bs can be distinguished from the T-28D by the smaller nose wheel used on the Navy aircraft.

PHOTOS TOP TO BOTTOM:

T-28B, 147734, in standard Navy trainer colors with Navy codes painted out and Air Force codes overpainted. Photo taken April 1970 by J. Wible via Roger Besecker.

T-28B, 158228, in Air Force overall grey colors. Note the Navy's white-red-white prop tip was retained. Photo by Baker via Menard.

T-28B, 140049, in the same markings as above. Photo by R. Esposito via R. Besecker.

T-28B, 147796, again in Air Force grey but with yellow prop tips. Photo taken on 23 Oct. 1970 by J. Wible via R. Besecker.

T-28B, 158277, grey with black fin stripe and fin tip. Photo from Lionel Paul collection.

T-28D

The T-28D came about by remanufacturing the T-28A aircraft stored at MASDC (military aircraft storage and disposal center) since 1959. Basically these aircraft were brought up to "B" standard. Additionally the T-28D wings were strengthened to hold six hardpoints instead of the normal two. Armament specifications and stores locations are the same for the T-28D as for the modified T-28C, outlined later in the text.

The exact number of T-28Ds manufactured is not known, nor is their ultimate destination.

THE T-28D AND VIET NAM

In early 1961 the U.S. Army Special Forces entered Viet-Nam to train the South Vietnamese. Based on this lead the USAF started activating its first special unit for guerrilla warfare since WWII. These units would become known as "Air Commandos". The first Air Commando unit, established on 14 April 1961, was the 4400th Combat Crew Training Squadron based at Eglin AFB. The unit's original allotment of aircraft were 16 C-47s, 8 B-26s and 8 T-28s. The purpose of this unit, like the Army Special Forces was to train the South Vietnamese. In this capacity, detachment 2A of the 4400th left for South Vietnam on 11 Oct. 1961. For combat purposes, the 4400th was code-named Jungle Jim, and detachment 2A was called Farm Gate. Farm Gate arrived in country with 8 T-28s, 4 SC-47s, 4 B-26s and 151 men. Additionally, all aircraft carried VNAF markings.

Being a training unit for the South Vietnamese, Farm Gate's early combat directive was to "fire back if fired upon". This directive was augmented by one dated 26 Dec. 1961, which required a VNAF crewman aboard during combat operations. T-28s were also authorized to aid the VNAF in emergency situations when the VNAF could not perform the mission. So, like its Army counterparts, the Air Commandos found themselves in a limited combat capacity by the end of 1962.

In addition to Farm Gate's T-28s, the South Vietnamese received 30 T-28s on 13 Nov. 1961. This allotment would allow the VNAF to form its first T-28 squadron, known as the 2nd fighter squadron. In Feb. 1962 Farm Gate received five replacement T-28s plus an additional five in Jan. 1963. 1962 would see another eighteen T-28s shipped to Nam to form the nucleus of a new squadron. The squadron, established in Sept. 1962, was known as the 716th recon squadron, and was allotted 18 RT-28s and 3 RC-47s. A RT-28 was simply a T-28 fitted with a special photographic belly pack. Some of the T-28s used by this squadron were modified Navy T-28Cs with the tailhook removed.

June 1963 marked the end of Farm Gate. The aircraft and personnel of the de-commissioned unit became the 34th tactical group. Under this new structure the 1st Air Commando squadron was activated. During 1963 the Air Commandos would encounter increased ground fire and the resulting losses. In Sept. 1963, .50-caliber ground fire would claim one T-28 during the heavy fighting around Soc Trang airstrip. Then on 19 Oct. 1963 six T-28s were damaged by ground fire while defending the ARVN 21st division. Another T-28 was lost and one was damaged when on 24 Nov. 1963 a mission was attacked by hidden enemy forces equipped with five 7.9 mm guns and a twin .50-caliber weapon. 1964 brought not only more losses by ground fire, but something worse — structural failure. Structural failure in the form of wing separations claimed the lives of two men, Capt. Edwin G. Shank Jr. on 24 March 1964, and Capt. Robert Brumett on 9 April 1964. Because the appearance of North Vietnamese regulars and the heavier weapons they fielded made ground attacks in T-28s very unhealthy, and the fact that structural failure was emerging, the Air Force withdrew the T-28 in 1964-65. The Air Force replaced the Trojan with A1-Es in the Air Commando squadrons and with A1-Hs in VNAF units.

Although the T-28 was replaced in Viet Nam, it would appear again in strength in the Royal Laotian Air Force in 1966. As early as April of 1964, Detachment 6 of the 1st Air Commando wing, stationed at Udorn, Thailand, was responsible for checking out Laotian pilots in T-28 aircraft. By 1967 the unit's strength had risen to approximately fifty T-28s.

PHOTO AT LEFT: Factory fresh T-28D, 0-17665, in the natural metal and grey scheme. Note absence of USAF markings. Prop is silver with back ¾ of blades black. Gun pod is natural metal and bomb pylons are white. Note gunsight in windscreen and the addition of armor plate to the seat backs. Photo by R. Miskesh via R. Besecker.

AD

13605

USAF
0-91558

U.S. AIR FORCE

PHOTOS AT LEFT TOP TO BOTTOM:

Two T-28Ds at Hulbert Field Fla., 51-595 is in the foreground. Note how exhaust has washed away the camouflage. On the camouflaged T-28D the props are all black with yellow tips. These aircraft have what is believed to be a red stripe on the tail fin and a tail badge. Also note the complete tail code is white. The most interesting feature of these two aircraft is the fact that ejection seats have been installed. Note the warning triangles under the canopies. Photo from Lionel Paul collection.

T-28D, 51-3605, tail number is white with small USAF in black painted above it. Top of fin tip is light grey. Photo by Steve Miller via Menard.

T-28D, 0-91558, with large fuel tank and white rocket container. Photo taken on 5 Nov. 1970 by Ira Ward via AAHS.

T-28D, 53-8370, in Air Force grey scheme and silver prop. Note very unusual fuel tanks and crude multiple ejector bomb rack. Photo by K. Kistler via AAHS.

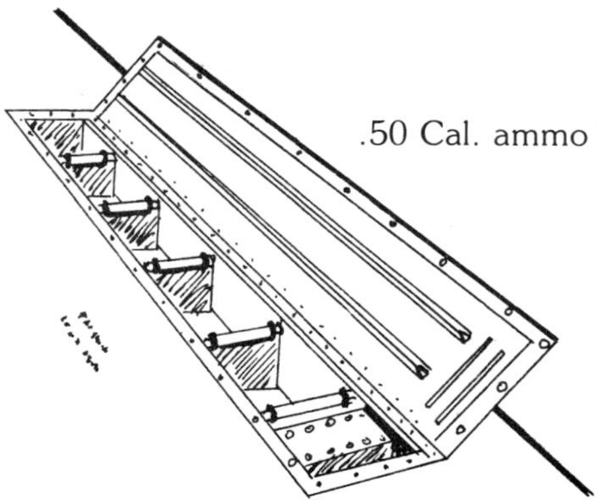

.50 Cal. ammo bay (Author)

PHOTO BELOW: T-28D, 0-17870, also in factory fresh color scheme. Photo by P. Bowers via R. Besecker.

PHOTO BELOW: T-28D, 51-480, in camouflage scheme of two colors of green and tan with light grey under surfaces. See back cover. AD on tail stands for the 4407th. CCTS. The only reference to Air Force is in the tail markings. "AF" is painted in small black letters in front of the white 480 and above the number 51. Note large stores carried on pylons. Photo by Paul Stevens via R. Besecker.

PHOTO AT LEFT: T-28D, 0-13787, in all-over grey color scheme and no national markings. This is the way all T-28s left the U.S. for Nam in the early years. Photo via AAHS.

PHOTO BELOW: T-28D in all grey paint scheme and no rational markings. Note gun pod, napalm, 500 lb. bomb, and rocket pod under each wing.

PHOTO ABOVE: T-28B of the 4407 CCTS, 1st. SOW, in the green and tan scheme. Air Force photo via Lionel Paul.

T-28D, 49-1591, of the 606th SOS, Special Operations Squadron. Overall color, light grey with dark grey anti-glare panel and black prop with yellow tips. Exhaust area black,

T-28D, of the 4407 CTS, note tail code has changed from AD to AH. Also note the aerial torpedo under the wing. Photo, Lionel Paul collection.

T-28D, 51-605, 8 June 1969. Photo by Steve Miller via Menard,

PHOTO AT LEFT: South Vietnamese modified T-28C, 146244, with tail hook removed and larger nose tire and wheel added for rough field operations. This aircraft belongs to the 716th recon squadron and carries the unit's badge on the tail. Also note that camera package has been fitted. Overall color is grey upper-surfaces and white under-surfaces. Prop has a yellow dome and gun pod is silver. Observe how exhaust has almost obliterated the stars and bars. Bar on tail is yellow with three red stripes. Note squadron code of MO aft of star. Star and bars are white star on a blue circle surrounded by red; color between bars are yellow. Note whip antennae aft of canopy. Photo by David Menard at Da Nang, May 1965.

T-28 Camera Installation (Author)

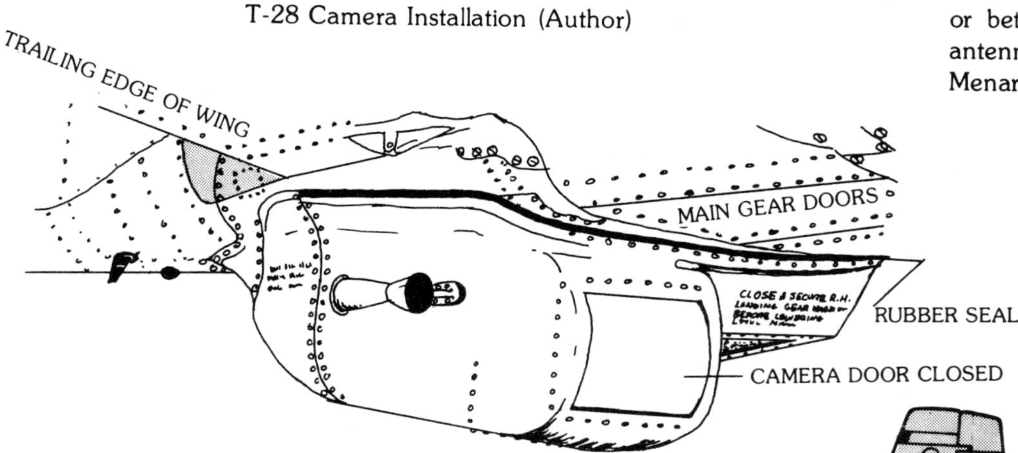

VNAF T-28C, 140457, note engine cowling strake.

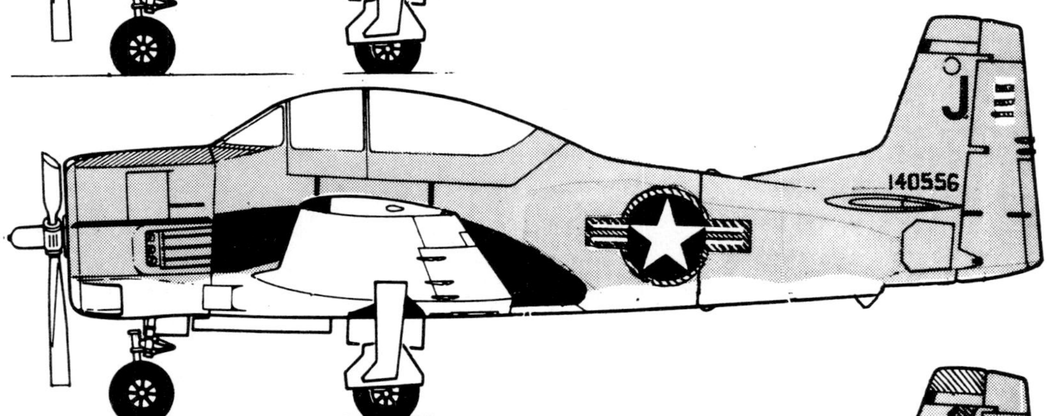

VNAF T-28C, 140556.

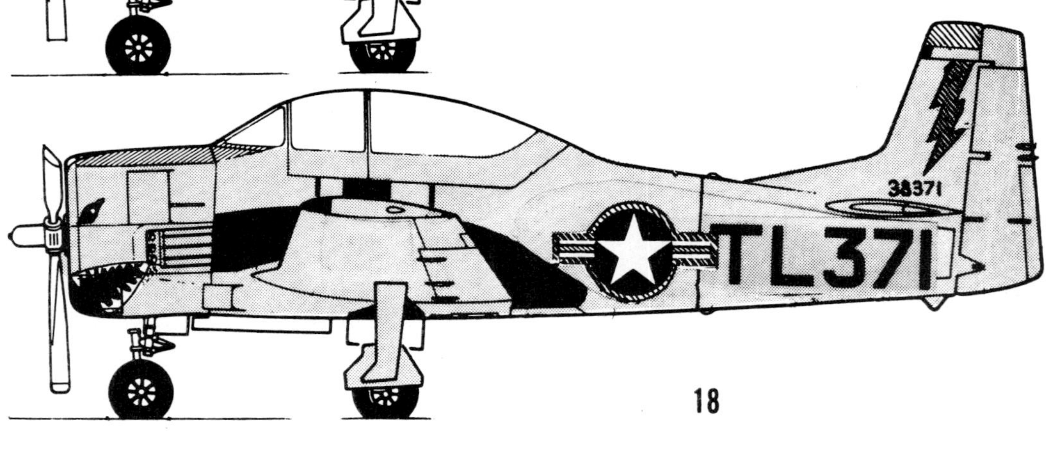

Alternate color scheme offered on the Heller model kit, T-28D, 38371, 1st Air Commandos. Red fin tip and lightning bolt and red mouth as on photo on back cover.

THE MONOGRAM T-28D, T-28B KIT

Although scarce, Monograms 1/48 scale T-28 kit makes a welcome addition to any collection. Originally produced in the 1950s as a T-28A, the kit does suffer from two minor problems. All the kits landing gear doors must be replaced as well as the wheels. Secondly, the cockpit area needs a lot of help. Outside of these two areas, the kit is really excellent.

The T-28D kit of the 60s-70s comes with all the necessary stores and a decal sheet depicting a VNAF T-28D.

The T-28B kit of the late 70s depicts an aircraft of VA-122 in trainer aircraft markings in effect at that time.

THE FENNEC OR DESERT RAT

In 1958, North American modified a T-28A into the company's NA-260 Nomad. The Nomad was a general purpose aircraft for civilian use. Pacific Airmotive Corp. of Burbank, Calif., was given modification rights and produced the Nomad Mk. 1 with a 1,300 h.p. engine and the Nomad Mk. 2 with a 1,425 h.p. engine. At this time France was looking for an inexpensive counter-insurgency aircraft. So France struck a deal on 245 surplus T-28As. Pacific Airmotive (PAC-AERO), did the modifications and Sud of France fitted the cockpits to French standards and provided armour plate for the crew. These aircraft were converted in 1960 and 1961 and used initially in Algeria. Eventually the Fennec would turn up in hotspots throughout the world.

HELLER T-28 FENNEC

The Heller T-28 is moulded in silver plastic and is flash free. On our sample there was one sink mark on top of the engine cowling. The rivet detail may seem excessive to some, but many large rivets were definitely used on the T-28. However, the access panel lines are etched too deeply and are what you would expect from Matchbox, but not Heller. It was disappointing to see that the canopy was not in three pieces; however, the panel lines are deeply cut and the canopy can be easily cut to build the model with the cockpit open. The most negative feature of the kit was the fact that the gun pods were not included. The wings are moulded with the fairings for the gun pods, and one stores station with a rocket pod for each wing. The decal sheet contains markings for a Fennec of E.A.L.A. 2/72 squadron French Air Force and a T-28D of the South Vietnamese Air Force.

BUILDING THE FENNEC

Our review kit was built up as the French Fennec. The construction of the kit was straightforward according to the directions. It should be noted that the nose has to be virtually packed solid with weights for the T-28 to sit on its landing gear. The only filling that was required was around the engine cowling. In order to give the kit proper detail, one must add braces from stretched sprue to the landing gear doors and static lines to the wings and tail surfaces.

If you want to build the kit as the VNAF T-28D you will have to cut the air scoop off the top of the engine cowling and fill it in. The big problem is that you will have to either cut down the canopy or make a new one as all T-28Ds were remanufactured without the rollover brace and were built with the lower canopy. Also, if you intend to build the kit as a Navy T-28B or C, you will have to replace the nose wheel with the smaller type used by the Navy. Building the kit as a T-28C will also require the cutting away or notching of the rudder section aft of the tail skid and adding a tailhook. If you opt for building the T-28A you can use the kit's canopy but will have to do major surgery to the nose section. At any rate the Heller kit is an excellent kit, and we say thank you, Heller, for a long-awaited kit in 1/72 scale.

PHOTO BELOW: French Fennec brought to America, N14141, ex USAF 51-3742, ex Fennec #104. Photo by Jerrold Narland via AAHS, taken 30 Jan. 1973. Note the major difference between the Fennec and T-28D or T-28B is the air scoop in front of the windscreen. Also note all French T-28s were made from the T-28As built with rollover brace between the front and rear cockpits.

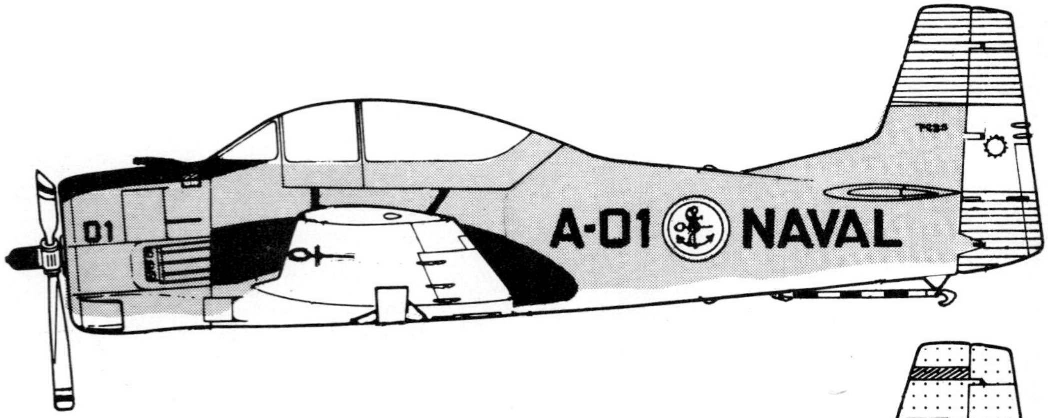

Argentine T-28S Fennec. Uppersurfaces light grey, lower surfaces and middle portion of the rudder white. Powder blue upper fin and lower rudder. Yellow circle on rudder appears to be a stylized sun. Black prop dome, anti-glare panel and exhaust stain area. Prop blades silver with white — red-white tips. Tailhook white and black stripes. National insignia is a black anchor with a yellow dot in the center. The object to the left of the anchor appears to be a red hat. The anchor is on a white field surrounded by two powder blue circles.

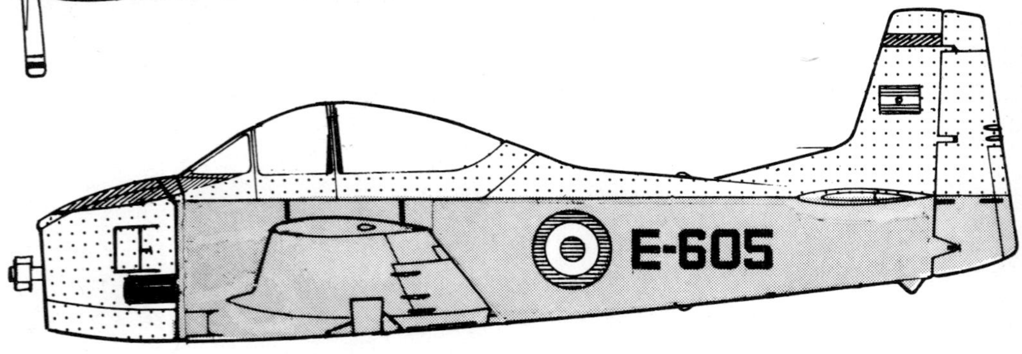

Argentine T-28A with natural metal upper surfaces and light grey lower surfaces. Dark grey anti-glare panel and fin stripe. Roundel powder blue -white - powder blue. Fin mark the same as roundel with yellow mark in the center.

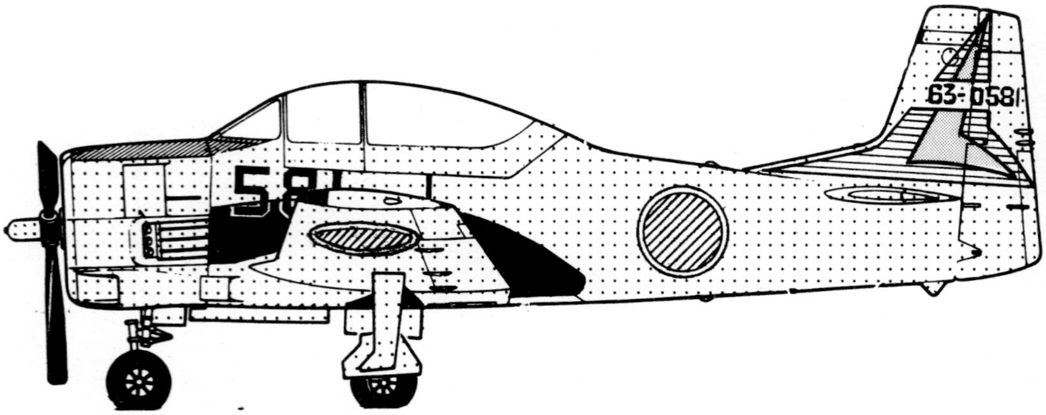

T-28B of the 501st. Photo Recon. Squadron, Japanese Air Self Defense Force. Overall natural metal with O.D. anti-glare panel. Fin markings blue with yellow center. Black prop.

Hamilton aircraft T-28R-1 of the Brazilian Navy, ex AF-50-202. Overall yellow white white fuselage bottom and white lettering. National insignia and tail stripes green-yellow-blue.

207 Sqn. Tan Indian head with brown hair and white horns on a green circle with an orange border. Feather is brown and white. War paint is red.

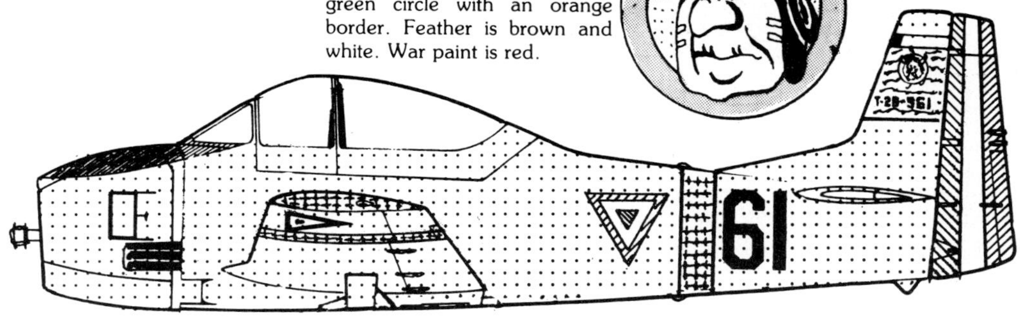

T-28A, 207 Squadron, 1st Air Group, Mexican Air Force, Ixtepec A.F.B., Oaxaca Mexico. Overall natural metal with O.D. anti-glare panel. Fuselage stripe, wing stripe and wing tip orange. Rudder markings from back, red-white-green. Squadron insignia on pinkish tan band. Triangular national insignia red - white - green.

T-28A, 201 Squadron Mexican Air Force, Cozumel A.F.B., Quintana Roo. Colors same as above except, wing stripe, fuselage strip and tail band yellow.

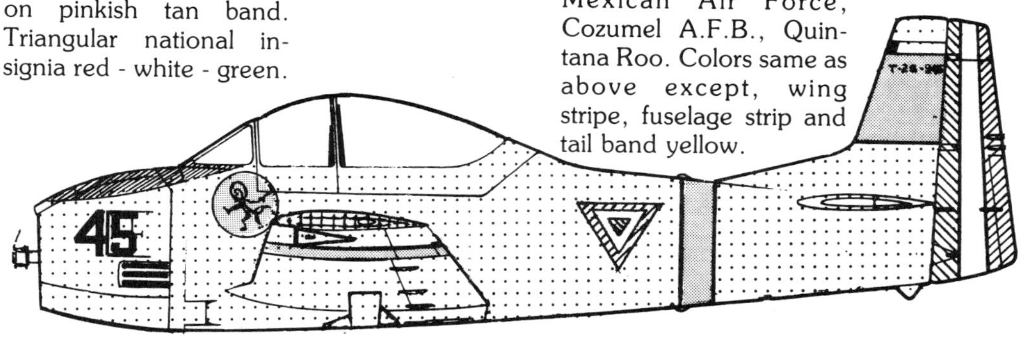

201 Sqn., Pancho Pistolas, blackbird with yellow beak and red mouth, dressed in green and brown on a yellow circle.

What follows is a list of all countries known to have used the T-28, where known applicable models are listed.

Argentina, T-28A and about 20 T-28s Fennecs fitted with tailhooks.
Bolivia, 6 T-28R1s remanufactured by Hamilton Aircraft Co. with tailhooks.
France, 245 T-28As remanufactured as Fennec, similar to D.
Cambodia
Castro's Cuba, 10 T-28As
Congo Republic
Dominican Republic
Equador
Ethiopia
Haiti
Japan, 1 T-28B
Laos, T-28D
Mexico, 30 T-28As
Morocco
Nicaragua
Rhodesia
South Korea, T-28A
South Vietnam, T-28D and C
Taiwan, T-28A
Thailand, T-28D
Tunisia
Zaire

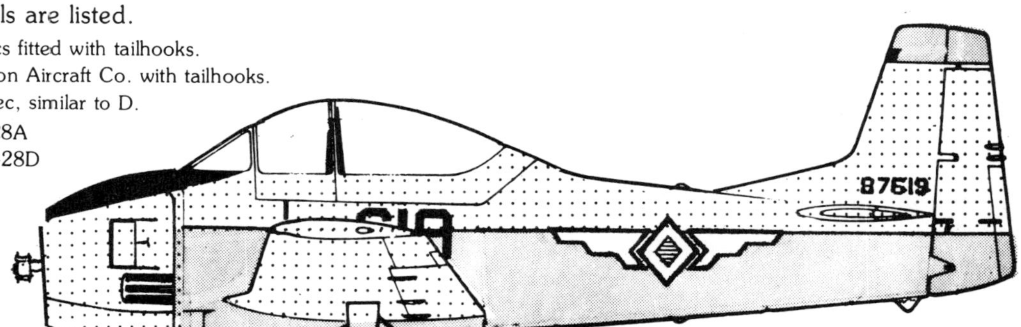

T28-A of the Philippines, overall natural metal with light grey lower fuselage. National insignia, white with center diamond and outline of wings blue, outline of outer diamond red.

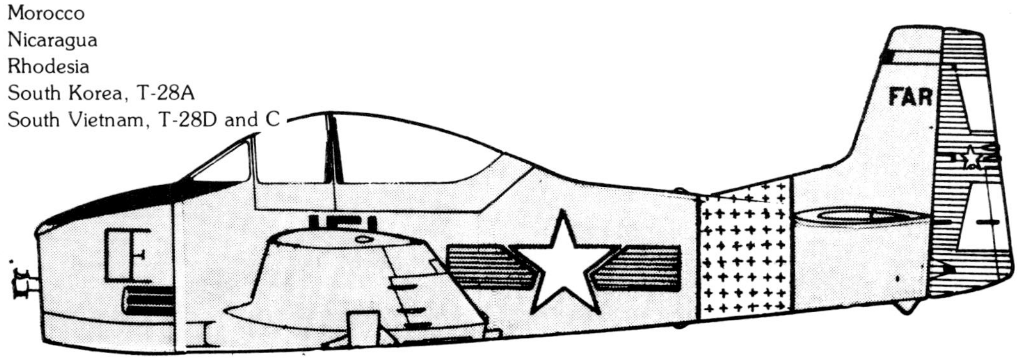

Cuban T-28A of the Fuerza Aerea Revolucionaria (FAR). Overall grey with orange fuselage band. Rudder is blue and white as is national insignia on fuselage side.

PHOTO ABOVE: T-28A, 51-7657, of the Korea Air Force. Overall natural metal with light grey lower half of fuselage and black anti-glare panel. Fuselage stripe believed to be yellow with red borders. Stylized red star on tail with three small white ones. Photo by Robert F. Dorr via AAHS.

PHOTO BELOW: T-28As of the Taiwan Air Force. Aircraft are overall natural metal with day-glo red nose, tail and outer wing panels. National insignia is white on blue, and nose gear wheel is black. Photo from the David Menard collection.

THE YAT-28E

Early in the South East Asian war, the Air Force realized it would need a coin aircraft more powerful than its remade T-28D. Therefore, in Sept. 1962, a $735,000 contract was awarded North American to modify a T-28A into the turboprop YAT-28E. The original designation was RA-28.

The first YAT-28E, 0-21242, took to the air on 15 Feb. 1963. On 27 March 1963, during its twelfth flight, YAT-28E number one entered an inverted flat spin and killed its pilot, George Hoskins.

After the crash the Air Force ordered two more test articles to continue the program. The second YAT-28E, 0-13786, was built with improved fittings and skins in the horizontal tail which increased the strength of these units by 250%. Two other refinements built into the number two plane was the elongation of the vertical stabilizer and the revised canopy which would allow usage of the new LW-2 ejection seat. First flight for the number two bird took place on 15 Nov. 1963. This aircraft was then put through a preliminary evaluation by the 1st Combat Applications Group, from 14 Feb. to 25 Feb. 1964. Further testing was carried out at Edwards AFB and then at Eglin AFB.

The number three YAT-28E, 0-13788, made its first flight on 30 June 1964. Number three differed from number two in having a new rear canopy which allowed installation of the LW-2 ejection system in the rear cockpit. Initial flight tests showed the canopy caused more control problems, so once again the vertical stabilizer was elongated.

In Aug. 1964 the Air Force suspended further testing and development of the YAT-28E even though North American had devised corrections for most of the aircraft's discrepancies. Although no specific reason was given for dropping the program, the fact is that the YAT-28E would not be the quick, simple and inexpensive fix the T-28D was.

The two YAT-28Es were put in storage by the manufacturer from Jan. 1965 till 24 Jan. 1966, when the number three aircraft was tested by the Navy. However, the Navy tested the aircraft as a trainer and not a combat aircraft. Navy testing was completed on 3 May 1966 with the conclusion that the YAT-28E was to much aircraft for fledgling aviators.

No production contracts were ever received for the YAT-28E as the Air Force bought the OV-10 Bronco and the Navy purchased the T-34C.

YAT-28E SPECIFICATIONS

The original 1,425 hp Wright piston-engine was replaced by a 2,450 shp Lycoming T55 turboprop, and provision was made for carrying up to 4,000 lb (1,815 kg) of weapons on 12 attachment points, including a Sidewinder air-to-air guided missile at each wingtip. The normal fuel capacity of 460 U.S. gallons could be supplemented by two 110 U.S. gallon underwing tanks.

DIMENSIONS:
Span 40 ft. 6 in.
Length 35 ft. 7 in.
Height 12 ft. 7 in.

WEIGHTS:
Weight empty 7,078 lb.
Max. T.O. weight 15,530 lb.

PERFORMANCE: (estimated)
Max. speed 360 mph
Max. cruising speed 276 mph
Econ. cruising speed 207 mph
Rate of climb at S/L 5,130 ft. min.
Max. range for ferrying 2,760 miles

PHOTO BELOW: YAT-28E, 0-13786, at Edwards AFB, May 1964. Note string prop brake used to hold prop from turning in the wind as it was free wheeling. Also note modified canopy, exhaust cover, wing test probe and belly radio antennae. Photo by D. Kasulka via R. Besecker.

PHOTO ABOVE: YAT-28E, 0-13788, colors as seen on back cover. This is the number three aircraft, note reshaped canopy and elongated tail fin. Photo by S. Miller via Menard, 24 Aug. 1966.

Entering Aircraft

These five drawings show the sequence used in entering the T-28 aircraft.

T-28B, VT-30, 130019, in overall white and red trim. Natural metal leading edges. Black prop and anti-glare panel. Yellow rescue arrow.

PHOTOS: top — T-28A, 137636, undergoing evaluation in early 1953. Note that aircraft is natural metal overall. Middle — Same aircraft on 11 June 1954 with the larger engine and three blade prop of the B model. Note the rivet pattern. National Archives Photos. Bottom — All blue T-28B, details of photo unknown. North American photo.

NAVY T-28B

In 1952 the Navy evaluated two Air Force T-28As (137636-137637). Soon thereafter on 6 April 1953 the first Navy T-28B took to the air. Subsequently, 489 T-28Bs were built.

T-28B Aircraft

Navair Numbers	Manufacture Sequence
137638 - 137810	1 - 173
138103 - 138367	174 - 438
140002 - 140052	439 - 489

T-28 AERODYNAMICS

The Navy T-28B and C Trojans are equipped with the R-1820-86 radial engine, manufactured by Lycoming. This nine-cylinder engine will develop 1425 BHP (brake horsepower) at 2700 r.p.m. (full increase) and 52.5″ MAP (manifold air pressure). The T-28B and C use a three-bladed, constant-speed, hydromatic propeller manufactured by Hamilton-standrd. The T-28B has a prop diameter of ten feet one inch, and the T-28C has a diameter of 9 feet 4 inches. The T-28C's prop diameter was reduced to prevent prop damage on arrested landings. To make up for the shorter prop, the C model's blades are wider than those on the T-28B and are stronger close into the hub. For this reason, the C model prop is less efficient than the T-28B prop and has a vibration range on the deck from 1900-2200 r.p.m.

The T-28B has a primary gross weight of 8,050 lbs. and the T-28C weighs in at 8,250 lbs. Useful loads for the T-28B and C are, crew 400 lbs., fuel 1,062 lbs., and engine oil 66 lbs. At these weights, maximum level speed for the T-28B is 270 kts. and 265 kts. for the T-28C.

The T-28 wing has both leading and trailing-edge taper. The leading-edge taper is approximately 2 degrees and the trailing-edge taper is approximately 8 degrees. This wing has a root chord of 8.2 ft. and a tip chord of 5.08 ft. The wing is also tapered in thickness towards the tip. This is done to improve the structural efficiency and to maintain a constant thickness ratio. This type of wing is suited for a square tip and, if the edges are blended, will reduce wing-tip vortices. The aspect ratio of the T-28 wing comes out to 6.078 with a wing area of 271.05 sq. ft. and a wing loading figure of 29.5 p.s.f. (pounds per square foot). Because the T-28 wing has a lower reynolds number at the tip, the wing is geometrically twisted 3 degrees to cause the root to stall first, instead of the tip. The T-28's wings are capable of lifting a maximum load of 12,370 lbs.

The thrust line of the T-28 is inclined 5 degrees. By inclining the thrust line, North American is able to reduce the destabilizing effect of power changes.

On the T-28B and C the speed brake is mounted on the underside of the fuselage, aft of the main landing gear. The operation of the speed brake is by an electrical switch on the throttle grip. The brake has only two positions — full up (off) or full down (on). Extending the speed brake reduces the pressure under the horizontal stabilizer. When the brake is extended, the nose of the plane pitches up. This pitch-up can be counteracted by forward stick force. The manufacturer places no restriction on speed brake use. The Navy has found that the violent nose-up pitch above 250 knots may cause the pilot to lose control of the aircraft. For this reason, the speed brake should not be extended above 250 knots.

SPEED BRAKE

The T-28 vertical stabilizer is offset one degree to the left. This aligns the vertical fin with the relative wind at design cruise airspeed and power.

The high-lift device on the T-28 is a single-slotted flap. High-energy air from the lower surface is ducted to the flap's upper surface. The high-energy air from the slot accelerates the upper surface boundary layer and delays airflow separation to some higher C_L. The slotted flap can cause much greater increases in $C_{L_{max}}$ than the plain or split flap and greatly reduce section drag.

An airplane will stall only when the critical angle of attack is exceeded. The critical angle of attack for the T-28 is 17 degrees (clean) and 14 degrees with the flaps down. To sustain balanced flight, THPa must be equal to or greater than THPr. THPa will vary with the brake horsepower, the propeller efficiency (np), and in the T-28, with exhaust thrust also. By adjusting manifold pressure, brake horsepower can be maintained at a constant value up to the critical altitude. Critical altitude is defined as the altitude, for your r.p.m., where maximum available manifold pressure equals maximum allowable manifold pressure. Items affecting stall speeds are listed in the chart below.

Stall Speed	Item	Stall Character
X	Abrupt Control Movement	X
	Altitude	
	CG Location	X
X	Coordination	X
	Cowl Flaps	X
X	Gross Weight	
	Landing Gear	
X	Power	X
	Speed Brake	X
X	Turns	X
X	Wing Flaps	X

The T-28 is restricted from doing the following maneuvers: (1) any maneuver that requires negative G's for over 10 seconds and (2) snap rolls.

The importance of doing proper stalls and adhering to aircraft G restrictions are graphically illustrated by the photo at right. This accident took place while practicing high speed stalls. It was concluded that this aircraft had been overstressed by at least 2 G's during the high speed stalls. This initiated the stabilizer/elevator failure. The pilot's failure to release back pressure at the onset of buffet in the 170-210 knot region and his subsequent abrupt forward motion of the stick when he felt a "pop" caused a rapid unloading of the high G forces on the aircraft. This caused the stabilizer/elevator to depart the aircraft in a fashion typical of a negative G overstress. The T-28B/C NATOPS manual states: "Because rolling pullouts impose additional stress, maximum permissible acceleration during rolling pullouts is 2/3 the maximum permissible acceleration for a normal pullout. At indicated airspeeds between 170 and 210 knots, critical combinations of acceleration and buffet can cause overstress. In clean configuration, the maximum permissible acceleration for flights in smooth air (dynamic stall line) is 5 G's at 170 kias and 6 G's at 210 kias, varying linearly from 170 to 210 knots. Stick back-pressure should be relaxed at the onset of the buffet in the 170-210 knot airspeed range.

In the lower fuselage, aft of the wing is a personnel cargo hatch, which folds down. This hatch weakens the structural integrity of the fuselage and is the main reason why snap rolls are prohibited. As late as 1971, students at Whiting Field were able to view the results of a snap roll firsthand. The T-28 used for bailout practice was a permanently grounded airframe. As a result of a snap roll, the rear fuselage and tail group were twisted a good 10 degrees by the overstress caused by the snap roll.

The T-28 glide ratio chart for all configurations follows:

	Knots Ias	**Glide Ratio**	**Rate of Sink**
Gear up, flaps up, speed brake and canopy closed	130	11.7 to 1	1120 f.p.m.
Gear down, flaps up, speed brake and canopy closed	110	8.1 to 1	1370 f.p.m.
Gear up, flaps down, speed brake and canopy closed	100	6.6 to 1	1525 f.p.m.
Gear down, flaps down, speed brake and canopy closed	100	5.7 to 1	1780 f.p.m.
Gear down, flaps down, speed brake closed, and canopy open	100	5.2 to 1	1940 f.p.m.

In a normal spin a T-28 will use about 500 ft. of altitude per turn. Spin recovery is accomplished in a T-28 by applying full opposite rudder with the stick slightly forward of neutral and neutral aileron. At stop of rotation the controls must be neutralized to prevent a progressive spin. A flat spin is caused by the center of gravity (CG) being moved aft of the neutral point and in the T-28 there is no recovery possible.

The maximum angle of bank for an aircraft is limited either by structural strength or power available. The maximum angle of bank for the T-28 in a level turn (at sea level) is 74° because of THPa.

Because rolling pullouts impose additional stress, maximum permissible acceleration is two-thirds the maximum permissible acceleration for a normal pullout. The maximum permissible

courtesy Approach

acceleration in a rolling pullout in smooth air for an 8050 lb. T-28 would be 4 G's (2/3 of 6) but only 2 2/3 (2/3 of 4) G's in moderate turbulence.

The Trojank's airspeed for flying in severe turbulence is 125 to 185 kts.

A conventional control system is used in the T-28. Proper control forces are produced by utilizing mass (static) and aerodynamic balance in conjunction with tabs, spring bungees, and a bobweight. To provide the desired degree of control-free stability, the control surfaces are balanced about the hinge line. Aerodynamic balance of a control refers to the effect of the air flow on the balance of the control surface. Aerodynamic balance is generally accomplished by placing part of the surface ahead of the hinge line. The type of aerodynamic balance used on the T-28's rudder is called overhang, while that of the elevator is the overhang design with the addition of a shielded horn. The mass balance of a control surface is generally accomplished by placing the necessary weight within the aerodynamic balance. In the T-28 rudder, lead weights are placed in the overhang (leading edge). The mass balance of the elevator uses a small weight in the shielded horn and a small weight in the overhang. The type of aerodynamic balance used on the T-28 rudder could be used on the ailerons, but at the high rotational velocity of which the T-28 is capable, aileron snatch could occur. So aerodynamic balance of the aileron is accomplished by the sealed internal method.

The T-28 has control gadgetry (feel) built into all three control systems. The purpose is to prevent the pilot from overcontrolling the airplane. The rudder uses a spring bungee (a cylinder with a spring inside) for feel and control centering. The elevator uses a spring bungee and a bobweight for feel and control centering. The bobweight also reduces the stick loading at high positive acceleration. The added feel for the ailerons is accomplished through the use of de-boost, or anti-servo tabs.

The T-28's aerodynamic center (A.C.) is located at 26¾% of the wing's chord. The C.G. travel limits for the T-28 are 18½% to 26¾% MAC. If the C.G. moves aft to the neutral point (N.P.), which is 36% MAC, the airplane would have neutral static stability (longitudinal).

The T-28 has positive directional stability, even with the destabilizing effect of power and "P" factor. "P" factor is the propeller yawing moment caused by one of the prope blades creating more thrust than the others.

The effective lateral stability in the T-28 is neutral. The dihedral angle of 8 degrees should give the aircraft positive static stability.

Landing Gear Details

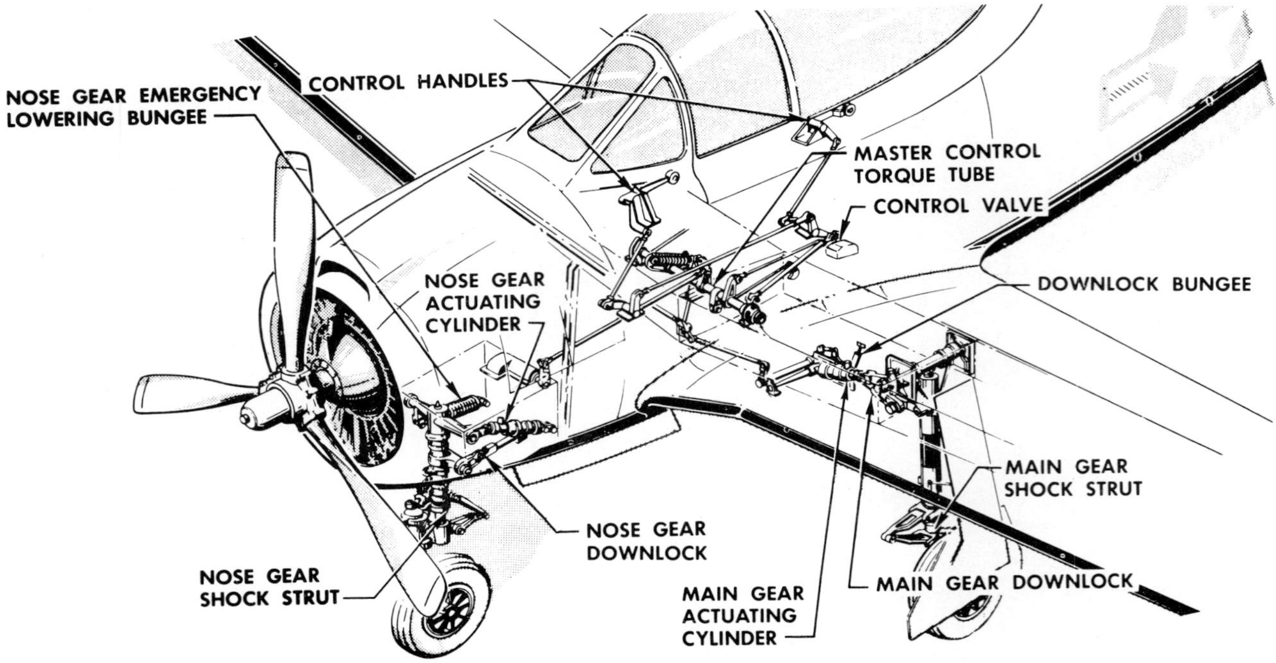

LANDING GEAR GROUND SAFETY LOCKS

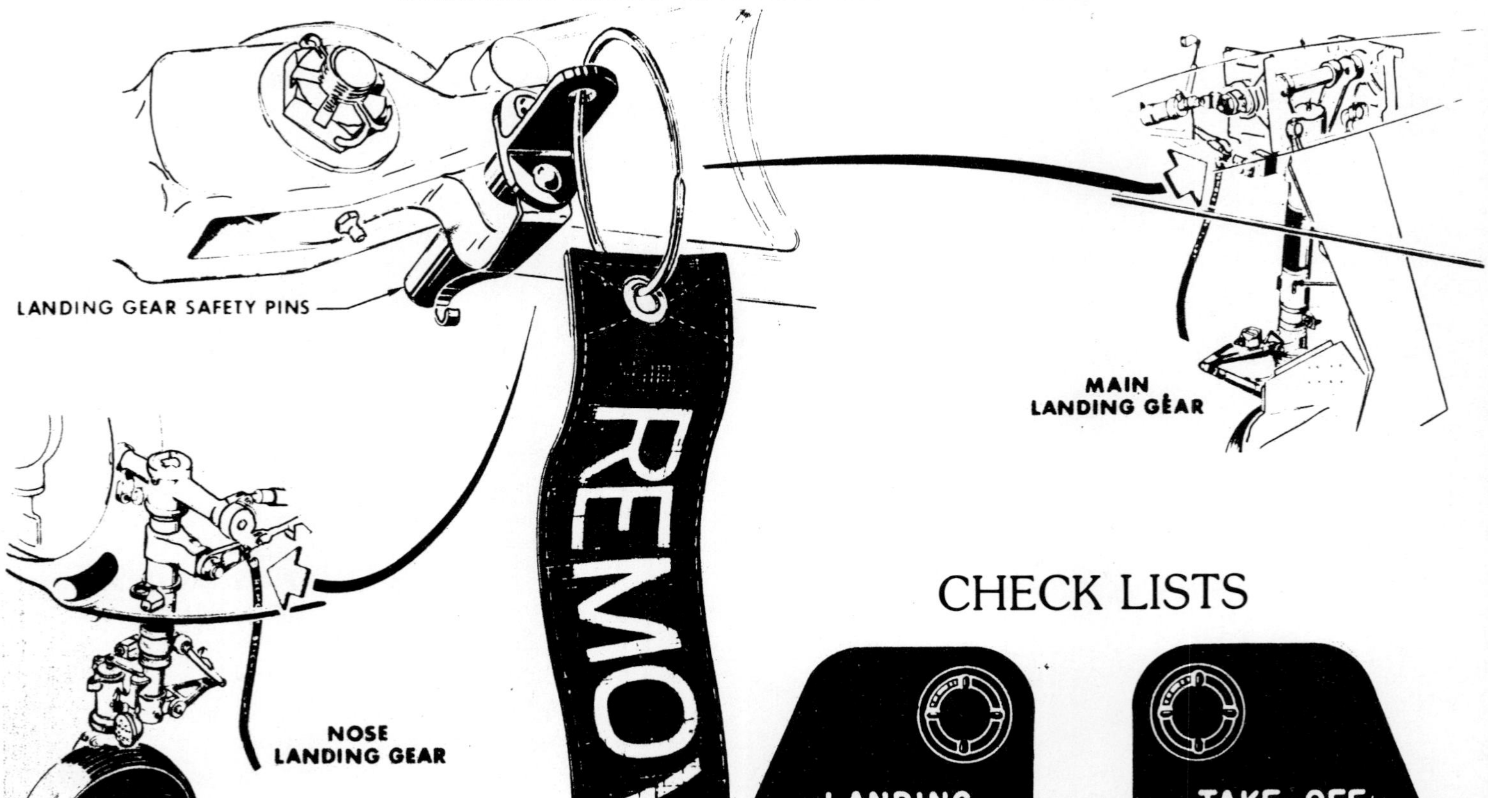

CHECK LISTS

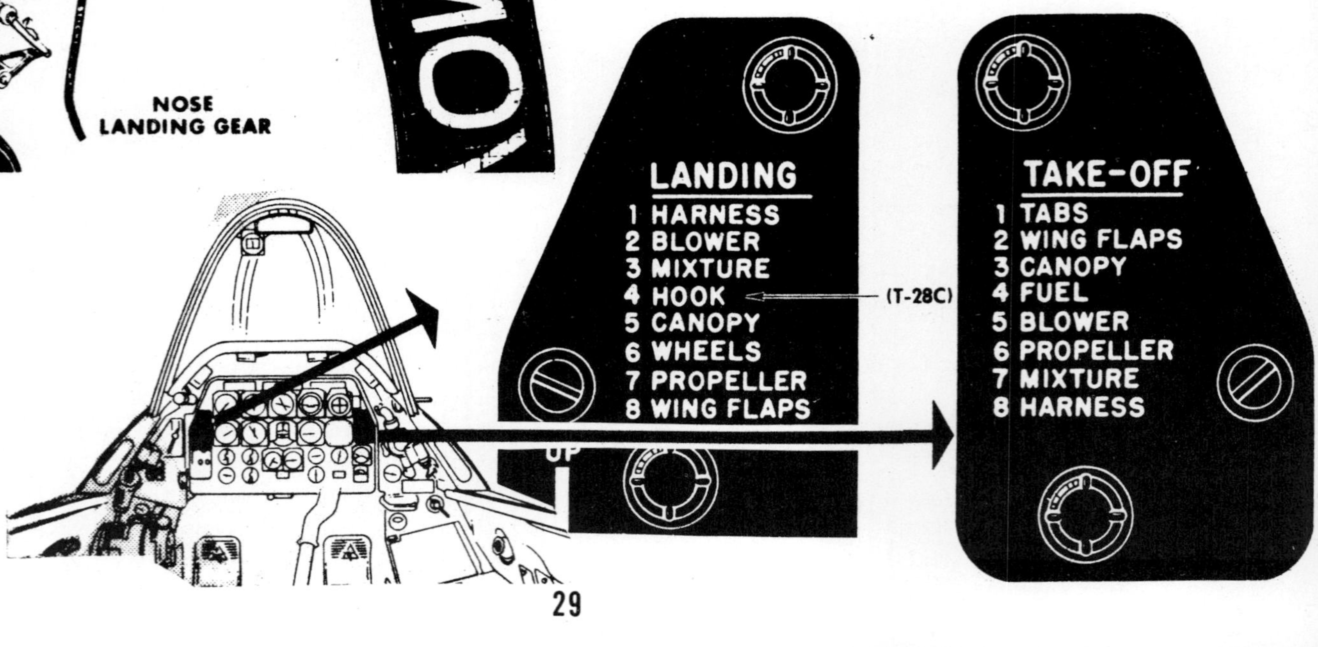

Instrument Panel

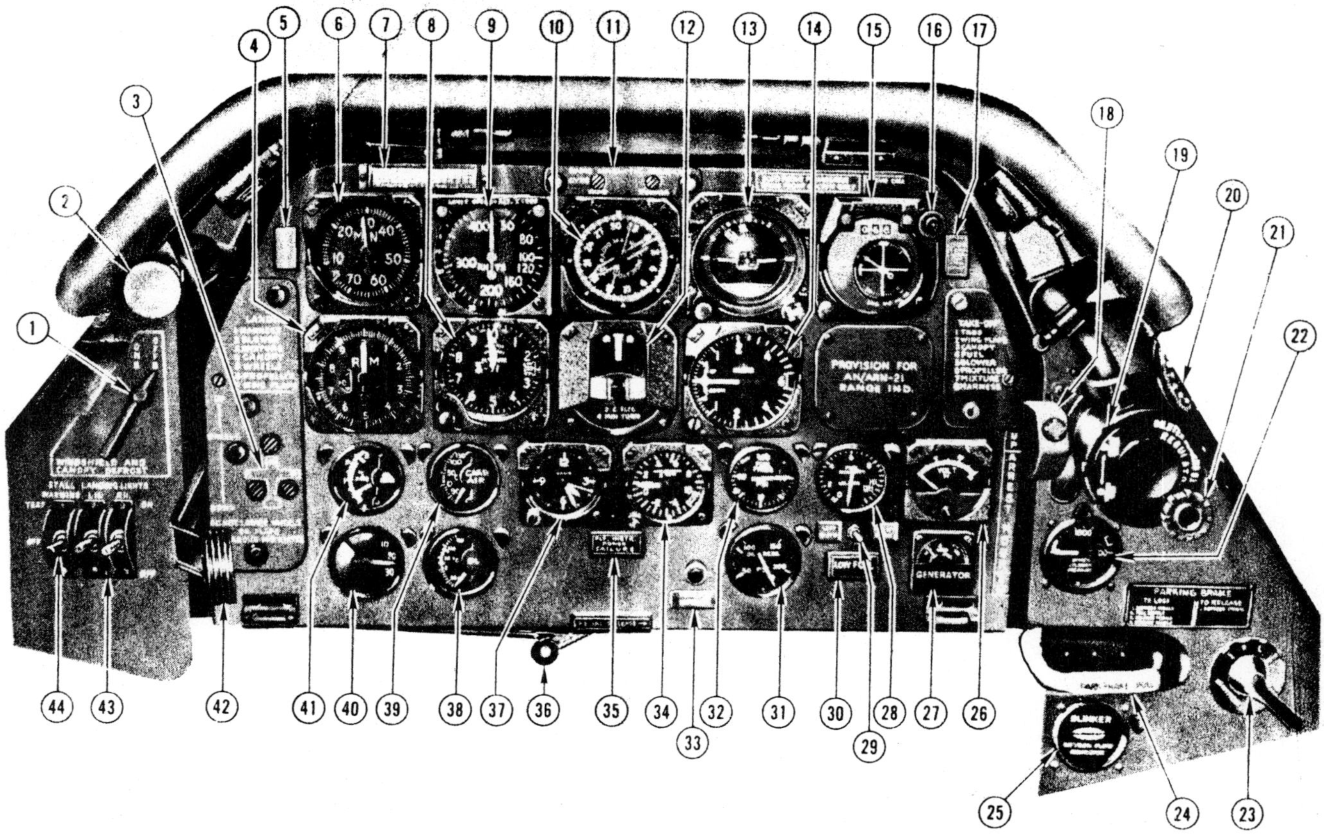

1. WINDSHIELD AND CANOPY DEFROST CONTROL HANDLE ◀
2. CANOPY EMERGENCY STOP BUTTON
3. LANDING GEAR POSITION INDICATORS
4. TACHOMETER
5. LANDING GEAR WARNING LIGHT (ASC 63)
6. MANIFOLD PRESSURE INDICATOR
7. AIRSPEED CORRECTION CARD
8. ALTIMETER
9. AIRSPEED INDICATOR
10. RADIO MAGNETIC INDICATOR
11. COMPASS ANNUNCIATOR
12. TURN-AND-BANK INDICATOR
13. GYRO HORIZON INDICATOR
14. VERTICAL SPEED INDICATOR
15. COURSE INDICATOR
16. MARKER BEACON LIGHT
17. COURSE SETTING CHANGE LIGHT
18. ARRESTING HOOK HANDLE (T-28C)
19. OXYGEN REGULATOR
20. FREE AIR TEMPERATURE INDICATOR ◀
21. OXYGEN REGULATOR EMERGENCY VALVE
22. OXYGEN PRESSURE INDICATOR
23. IGNITION SWITCH
24. PARKING BRAKE HANDLE ◀
25. OXYGEN FLOW INDICATOR
26. GENERATOR VOLTMETER
27. GENERATOR LOAD INDICATOR
28. FUEL QUANTITY INDICATOR
29. FUEL QUANTITY INDICATOR TEST SWITCH
30. FUEL LOW-LEVEL WARNING LIGHT
31. OIL PRESSURE INDICATOR
32. FUEL PRESSURE INDICATOR
33. SUMP PLUG WARNING LIGHT
34. ACCELEROMETER
35. FLIGHT INSTRUMENT POWER FAILURE WARNING LIGHT
36. RUDDER PEDAL RELEASE LEVER
37. CLOCK
38. OIL TEMPERATURE INDICATOR
39. CARBURETOR AIR TEMPERATURE INDICATOR
40. WING FLAP POSITION INDICATOR
41. CYLINDER HEAD TEMPERATURE INDICATOR
42. LANDING GEAR HANDLE
43. LANDING LIGHT SWITCHES
44. STALL WARNING TEST SWITCH (T-28C)

◀ FRONT COCKPIT ONLY

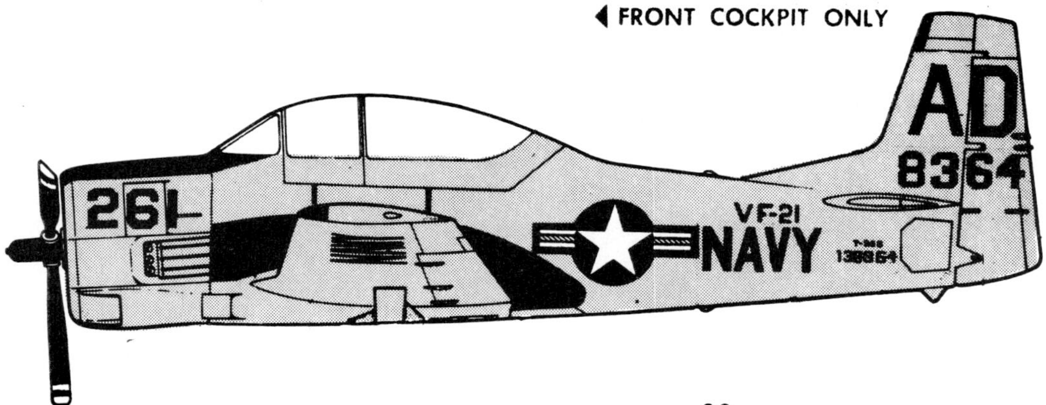

T-28B, 138364, of VF-21, in all yellow scheme with black lettering.

Left Console (Typical)

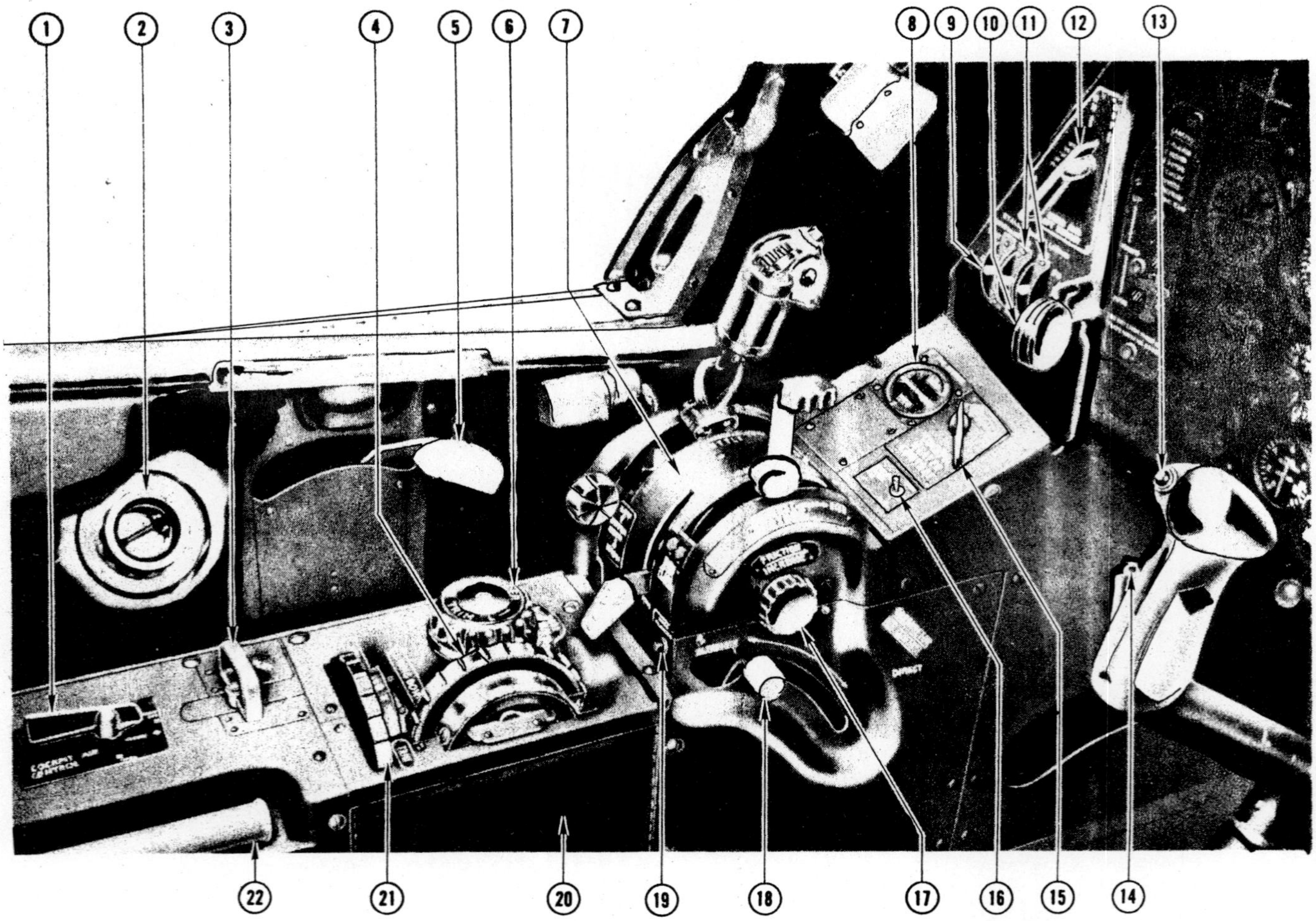

1. COCKPIT AIR CONTROL HANDLE
2. AIR OUTLET
3. FUEL SHUTOFF HANDLE
4. ELEVATOR TRIM WHEEL
5. CANOPY CONTROL HANDLE
6. RUDDER TRIM WHEEL
7. THROTTLE QUADRANT
8. HYDRAULIC SYSTEM PRESSURE GAGE
9. STALL WARNING TEST SWITCH (T-28C)
10. LANDING GEAR HANDLE
11. LANDING LIGHT SWITCHES
12. WINDSHIELD AND CANOPY DEFROST ◄ CONTROL HANDLE
13. BOMB RELEASE BUTTON ◄
14. ROCKET FIRING BUTTON ◄
15. COCKPIT HEATER CONTROL HANDLE ◄
16. COWL AND OIL COOLER FLAP SWITCH
17. FRICTION LOCK KNOB ◄
18. CARBURETOR AIR CONTROL
19. HORN SILENCER BUTTON (SOME AIRCRAFT)
20. FLIGHT REPORT HOLDER
21. AILERON TRIM WHEEL
22. HYDRAULIC HAND-PUMP ◄

◄ FRONT COCKPIT ONLY

Right Console (Typical)

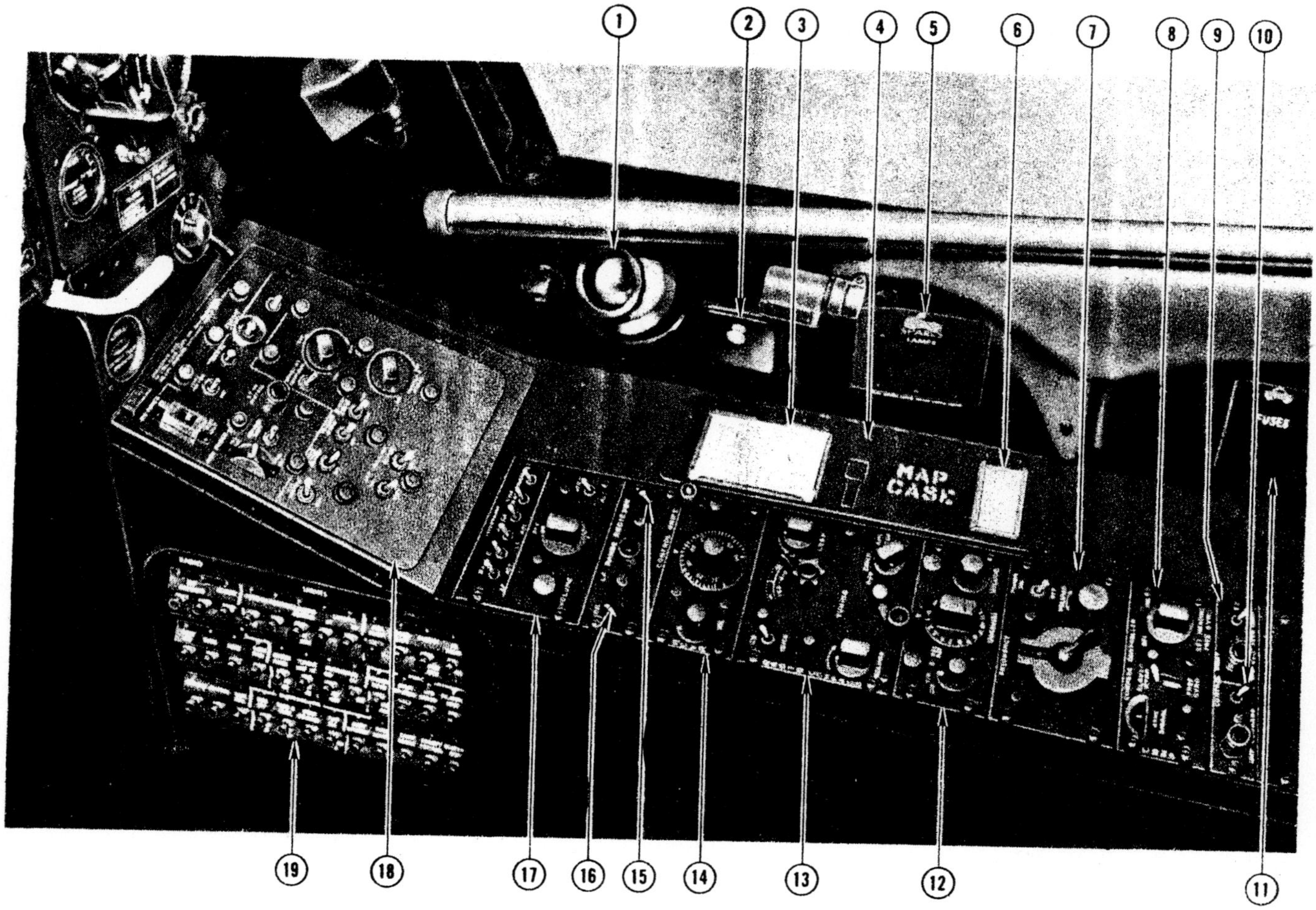

1. AIR OUTLET
2. ASH TRAY
3. CHANNELIZATION LOG
4. MAP CASE ◀
5. SPARE LAMPS ◀
6. COMPASS CORRECTION CARD
7. OMNIDIRECTIONAL RECEIVER
8. GYROCOMPASS CONTROL PANEL
9. NAV AND COMPASS CONTROL TRANSFER PANEL
10. COMMAND RADIO CONTROL SHIFT TRANSFER SWITCH
11. SPARE FUSES ◀
12. UHF RADIO CONTROL PANEL
13. RADIO COMPASS CONTROL PANEL
14. RANGE RECEIVER CONTROL PANEL
15. RANGE RECEIVER CONTROL SHIFT SWITCH
16. RANGE RECEIVER POWER SWITCH
17. INTERCOM CONTROL PANEL
18. ELECTRICAL SWITCH PANEL
19. CIRCUIT BREAKER PANEL ◀

◀ FRONT COCKPIT ONLY

NORTH AMERICAN AVIATION, INC.

T-28B

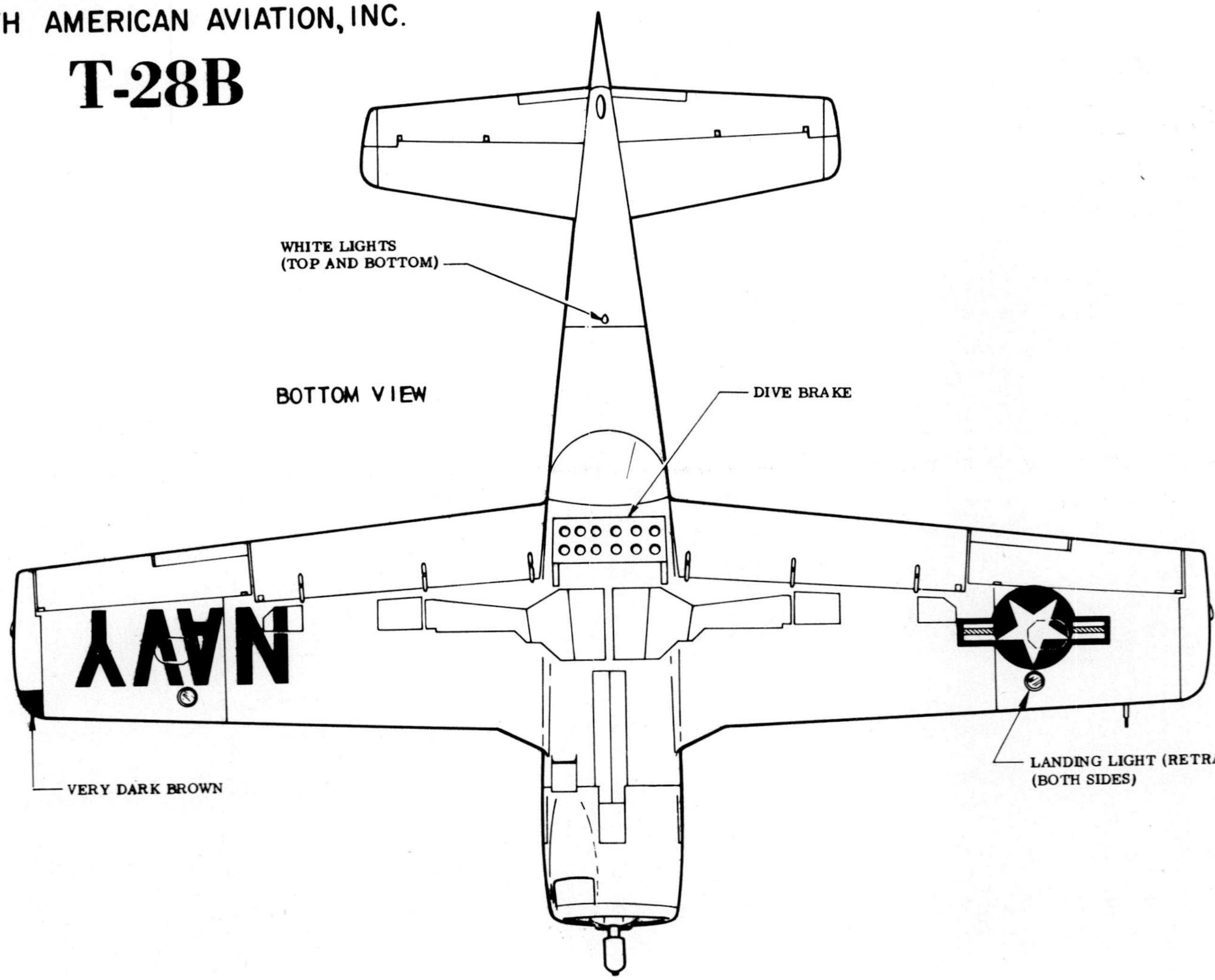

BOTTOM VIEW

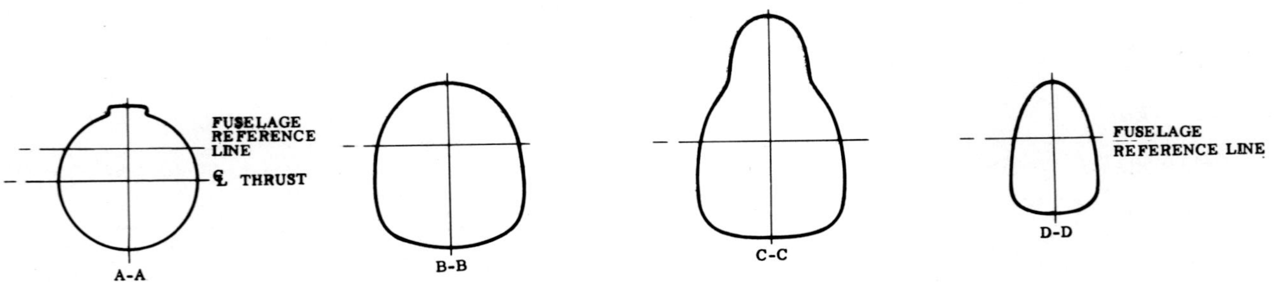

CROSS SECTIONS OF FUSELAGE

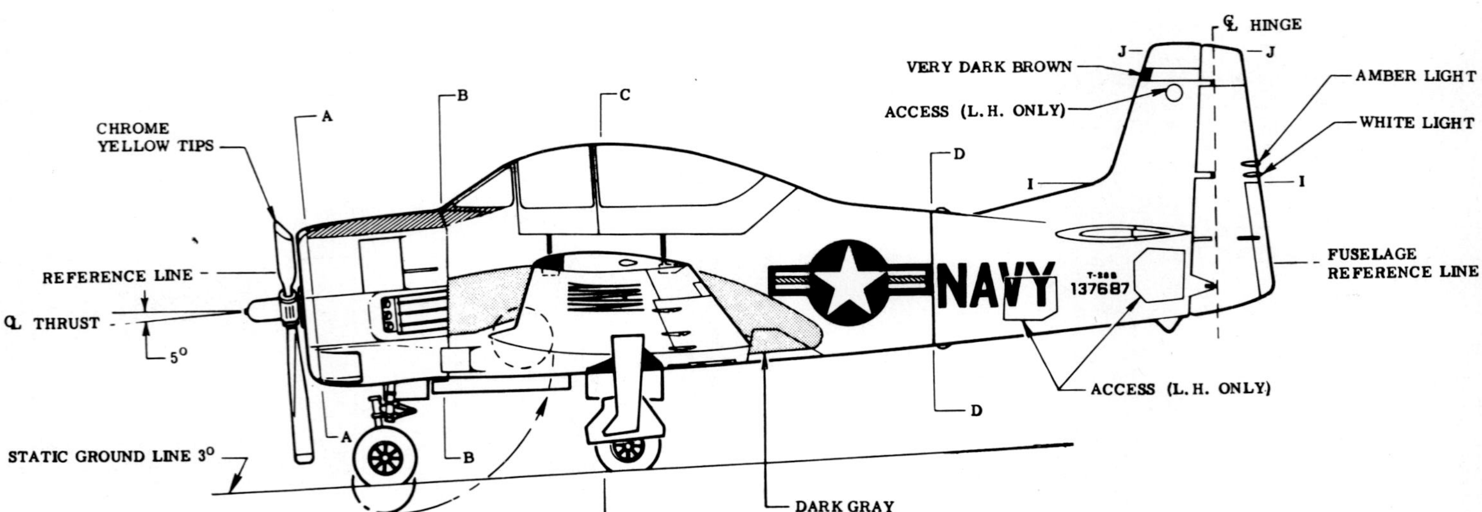

SIDE VIEW

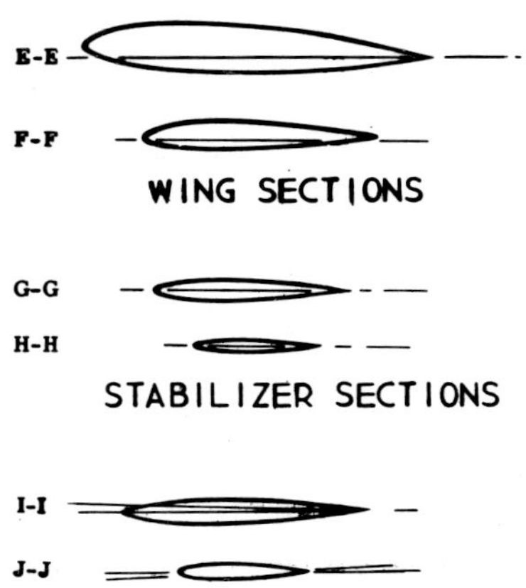

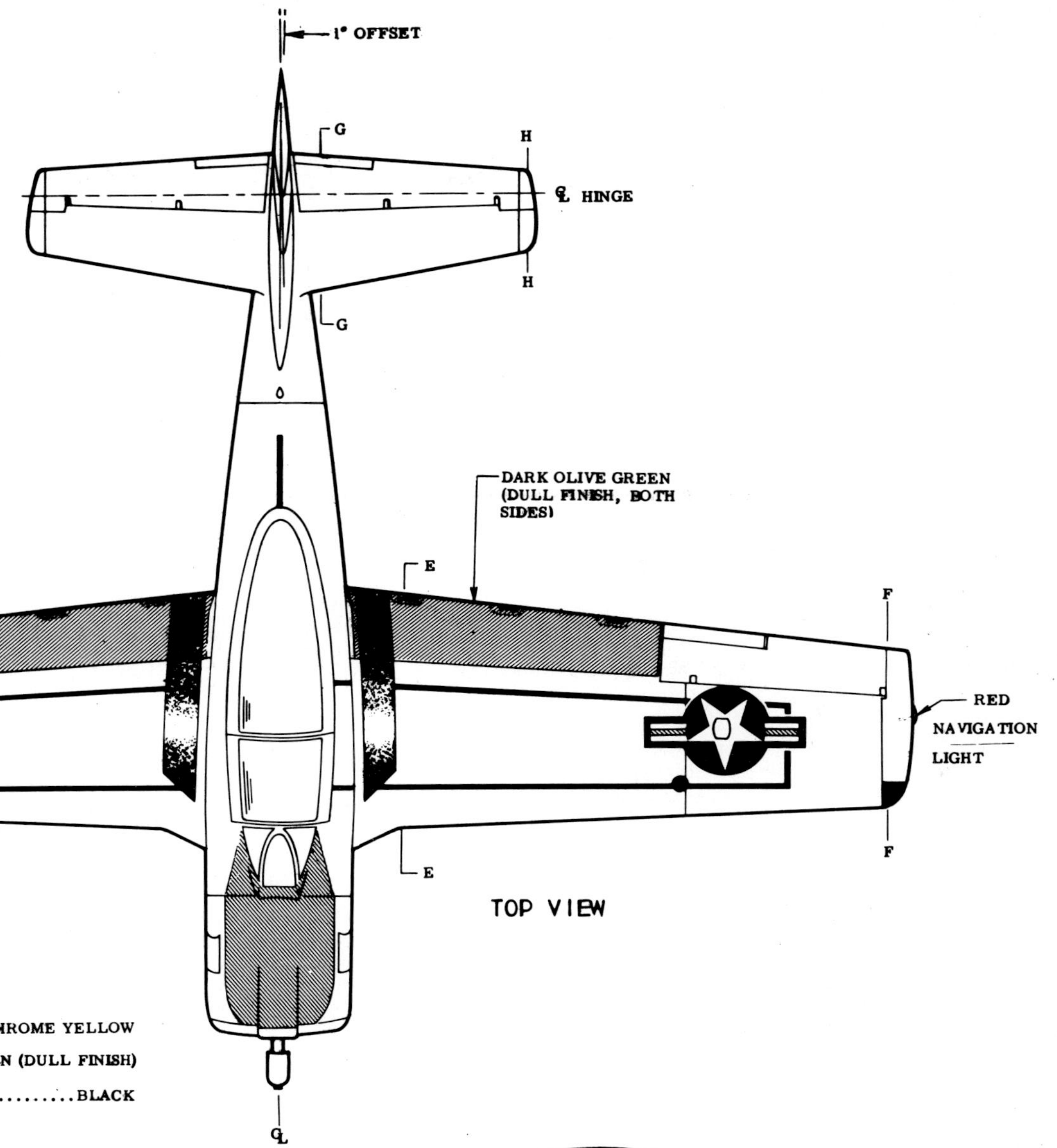

TOP VIEW

COLOR NOTES

DATE: MARCH 1954

ENTIRE PLANE.................................CHROME YELLOW

ANTIGLARE PANEL .... DARK OLIVE GREEN (DULL FINISH)

ALL LETTERING AND MARKINGS...............................BLACK
(UNLESS OTHERWISE INDICATED)

NATIONAL INSIGNIA.........................RED, WHITE, AND BLUE

WING SPAN.......................................40' 7-1/4"

LENGTH...32' 11"

HEIGHT...12' 8-1/2"

WING AREA.................................271.1 SQUARE FEET

POWER PLANT.......................WRIGHT 9 CYL. R-1820-86

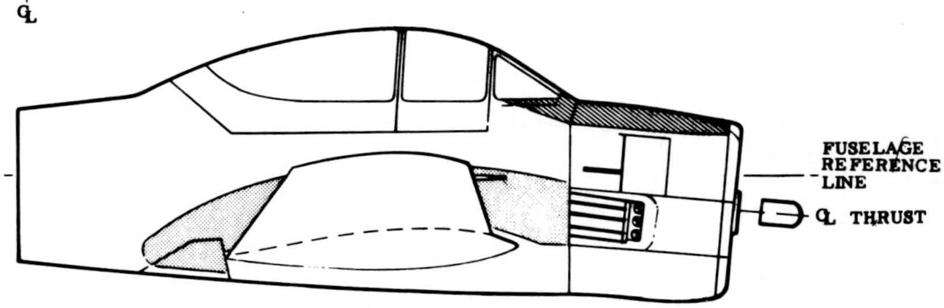

DETAIL OF RIGHT SIDE

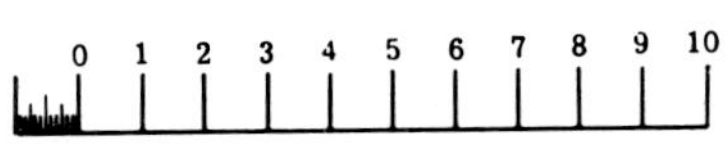

SCALE IN FEET

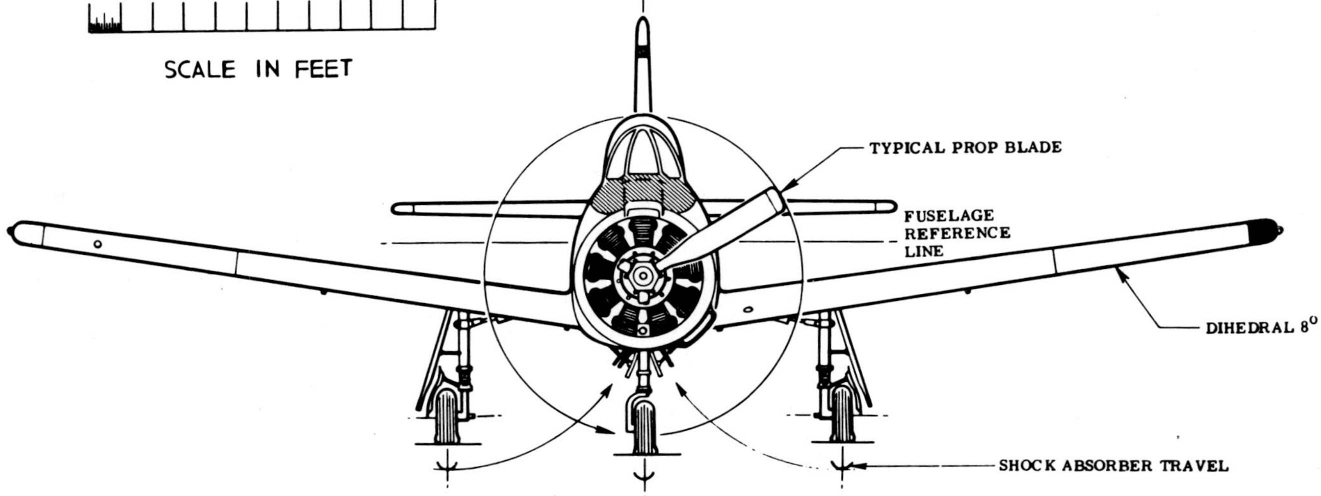

FRONT VIEW

EARLY NAVY T-28 TAIL CODES

When the T-28 first joined the fleet, it entered a world of yellow trainer aircraft and two letter tail codes. A few years later, in 1956-57, when trainer yellow was changed to white and red, tail codes were changed to one number and one letter. The early two letter tail codes are as follows:

PA — Instructors Basic Training Unit, based at Chevalier Field NAS Pensacola.
WC — Basic Training Unit One, based at North Whiting.
WB — Basic Training Unit One, based at South Whiting.
CA — Basic Training Unit Two, based at Corry Field.
SB — Basic Training Unit Three, based at Saufley Field.
BA — Basic Training Unit Four, based at Barin Field.
CD — ATU-800, All Weather Flight School, Corpus Christi.
KE — ATU-801.
SE — H & MS-10 USMC.
WD — Marine Instrument Training Squadron Ten.

PHOTOS: Above — T-28B aircraft of ATU-800, flying over their home base at Corpus Christi, Texas, in March 1955. National Archives Photo. Below — T-28B, 137692, of ATU—800. Note large Air Force type nose wheel that was soon replaced in service with the smaller type. Photo by Warren D. Shipp, via Menard. Note these aircraft are the same in color as the ATU—801 bird on the back cover, overall yellow with green fuselage and wing stripes.

T-28B, 137650, BTU—2, in overall yellow with black lettering and dark grey anti-glare panel.

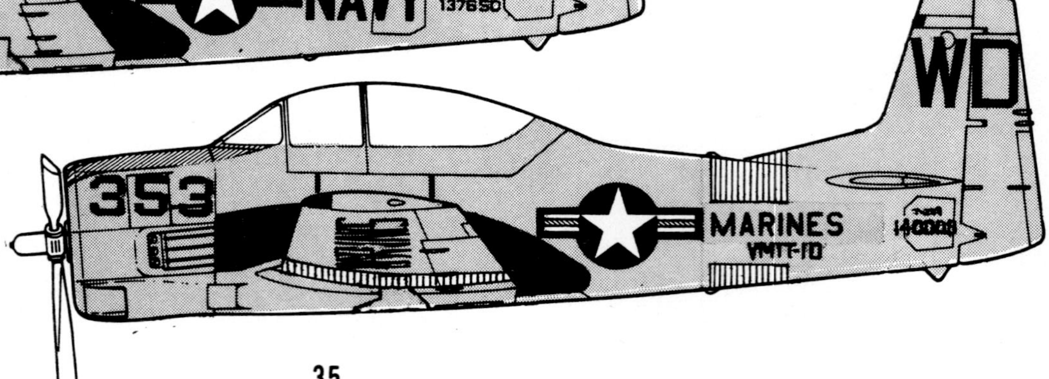

T-28B, 140008, VMIT—10, in overall yellow with green fuselage and wing stripes, 1956.

Service Access

❶ RIGHT WING FUEL TANK FILLER CAP (LH SAME)

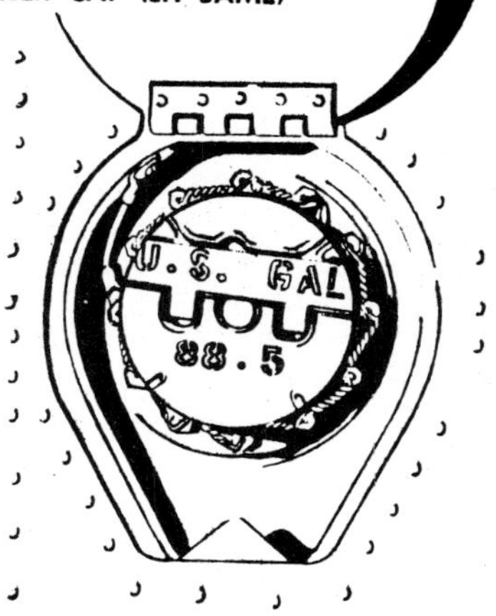

❷ OXYGEN AND FILLER ACCESS (BAGGAGE COMPARTMENT) T-28C

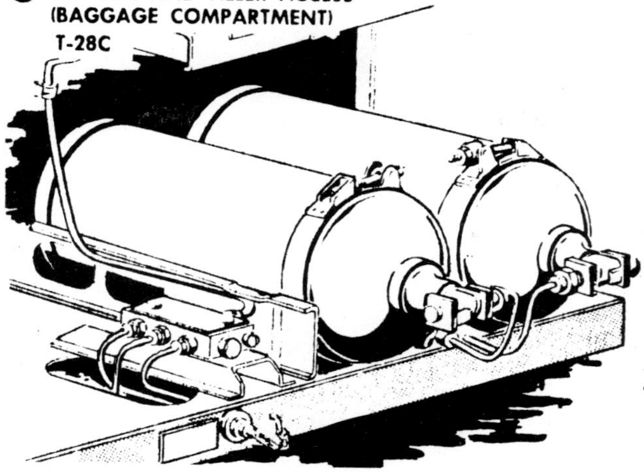

❸ CANOPY AIR BOTTLE FILLER ACCESS

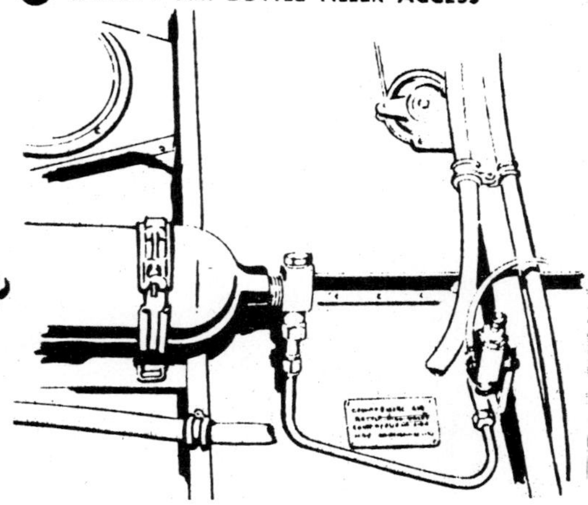

❹ ARRESTING HOOK SNUBBER ACCESS T-28C

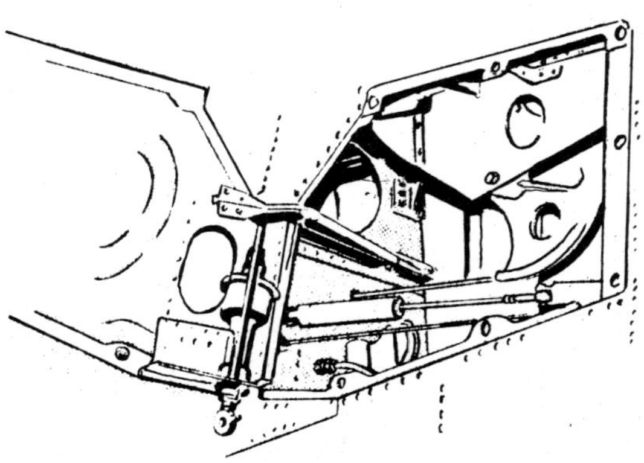

❺ BATTERY ACCESS

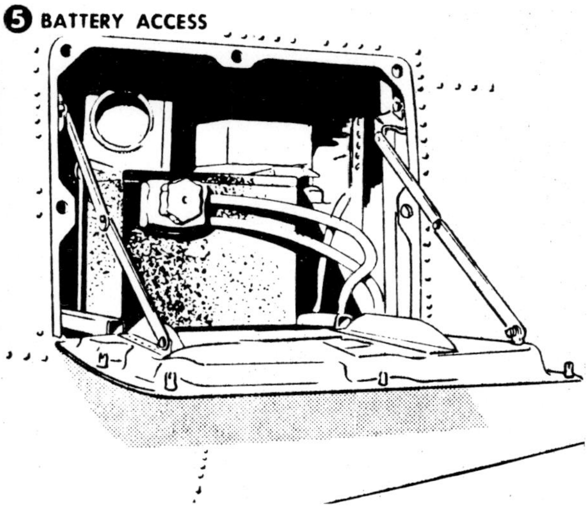

❼ HYDRAULIC RESERVOIR FILLER ACCESS

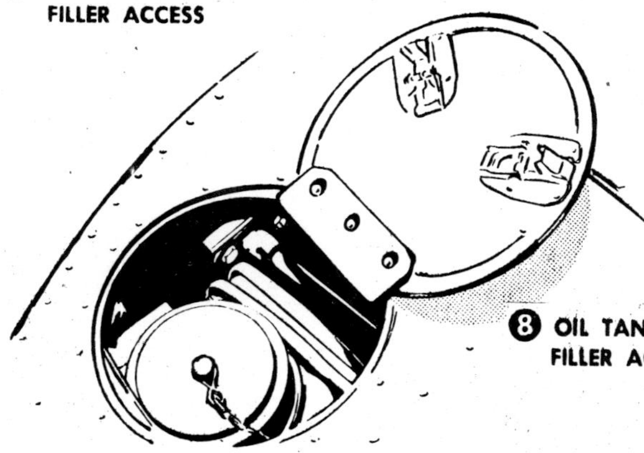

❽ OIL TANK FILLER ACCESS

❻ EXTERNAL POWER RECEPTACLE

PHOTOS AT LEFT TOP TO BOTTOM:

One of possibly two T-28As tested by Navy in the markings of ANACOSTIA. Natural metal overall with grey exhaust panel. Photo by Baker via Menard.

Series of four photos showing a factory fresh T-28B. Photo was taken in July 1955 and aircraft is painted in the all yellow scheme in effect at that time. Photos by Dave Menard.

T-28B, 138134, in the original trainer colors adopted in 1956. This Saufley station aircraft has the green tinted overhead on the canopy that many station aircraft use.

PHOTOS AT RIGHT TOP TO BOTTOM:

T-28B, 138273, station aircraft NAS Saufley. Late paint scheme, note white prop dome. Photo by Fred Roos, NF-R260-3, taken on 20 March 1971.

T-28B, 138236, station aircraft NAS Saufley in original yellow color with black lettering. Note green tinted canopy overhead and white landing gear. Photo by Ostrowsky via Menard.

T-28B, 138260, station aircraft NAS Whiting in all white paint scheme with ferry tanks installed. Photo by F. Hartman via R. Besecker.

PHOTO BELOW: T-28B, 137743, on 5 June 1971. Sword on tail and helmet on engine cowling are yellow. Prop dome is white. Photo by Harry B. Adams via AAHS.

SAUFLEY
8273
273
RESCUE
NAVY
T-28B
138273

SAUFLEY
236
236
NAVY

WHITING
8260

PHOTOS AT LEFT TOP TO BOTTOM:

T-28B, 138270, station aircraft NAS Whiting in all white color with natural metal leading edges. Photo taken on Oct. 1965 by A. Hafter via R. Besecker.

T-28B, 138358, in early red and white trainer scheme and green tinted canopy top. Note where 358 is painted under wing. NAS Pensacola station aircraft. Photo taken 26 Aug. 1962 by Paul Stevens via AAHS.

Another view of 138358, note natural metal leading edge on vertical tail and the way the white goes around the bottom of the star on the fuselage.

T-28B, 137662, Pensacola station aircraft in the late trainer scheme. Note that it is rare to see a T-28 parked without the flaps down because the steps are located in the flaps. Photo by Fred Roos, NF-R276-3, taken on 10 April 1971.

PHOTOS BELOW: Top — T-28B, 138363, station aircraft New Iberia. Compare size of man to T-28. Photo taken April 1964 by D. Kasulka via R. Besecker. Middle — T-28B, 137741, attached to the Naval Parachute Facility. Photo by Ostrowsky via Menard. Bottom — T-28B, 138166, station aircraft NAS Norfolk, note rescue other side sign. Photo by Fred Roos, NF-R302-8, taken on 2 June 1971.

PHOTOS AT LEFT TOP TO BOTTOM:

T-28B, 138326, in Sept. 1958 at the Navy's graduate school in Monterey, Calif. Color is all yellow with black lettering and red rescue arrow. Note landing gear is white. Photo by R. Dorr via R. Besecker.

T-28B, 137743, in early red and white scheme, Monterey, March 1961. Photo by D. Olsen via R. Besecker.

Two views of T-28B, 137741, used by the Chutting Stars in 1963. Aircraft is overall blue with yellow trim and lettering. Note Chutting Star emblem on upper tail and North American emblem on lower rudder. "Lt. J. G. Erickson" is written under canopy. Photos by T. Curdy and P. Stevens via R. Besecker.

T-28B, 138313, with NAF Andrews Washington DC on fuselage spine. Note NAF Andrews under wing. Photo by Seely via Menard.

T-28B, 138123, NAF Andrews with Marines on tail, 18 May 1962. Note how faded red is and how there is a white block around 8123 and NAF Andrews on tail. Photo by Gelbudas via AAHS.

PHOTOS AT RIGHT TOP TO BOTTOM:

T-28B, 138119, NAF Washington. Quite a large fleet of T-28s was operated from this field as the T-28 was an inexpensive aircraft to maintain in which officers on duty in Washington could implement their flight time. Photo by Fred Roos, NF-R287-9, taken on 31 May 1971.

T-28B, 138207, station aircraft NAS Meridian. Note the faded appearance of the early red and white scheme. Photo by R. Mikesh via R. Besecker.

T-28B, 140006, in all white paint scheme with very faded day-glo tail design outlined by a thin black line. Photo by Fred Roos, NF-R773-9, taken on 14 Feb. 1976. Aircraft was stationed at NAF El Centro.

WASHINGTON
8119
NAVY
138119

207
MERIDIAN
8207
NAVY
138207

0006
NAVY
140006

PHOTOS AT LEFT TOP TO BOTTOM:
T-28B, 137698, station aircraft NAS Norfolk. Note size relationship of men to plane. Photo from Lionel Paul collection.

T-28B, 138239, from NAS Sanford. Note small 239 on front cowling and natural metal leading edges. Photo by R. Harrison, 1964.

T-28B, 13, from Test Pilots School at NAS Patuxent River Maryland. Note red TPS painted on a blue tail stripe. Photo from the Lionel Paul collection.

T-28B, 137652, also from the Test Pilots School. Note stylized black Skyray on the tail. Photo by R. Besecker, taken on 21 April 1971.

T-28B, 137652, with very faded day-glo. NATC aircraft with wingtip test prob. Photo taken on July 1965 by Jim Burridge.

PHOTOS AT RIGHT TOP TO BOTTOM:
T-28B, 140047, MCAS New River. Note black nose wheel. Photo by Fred Roos, taken on 6 Feb. 1971. NF-R593-10.

T-28B, 138134, station aircraft MCAS Beaufort. Photo taken on 8 Oct. 1973 by Fred Roos. NF—R250-11.

T-28B, 138173, MCAS Quantico. Note number 173 on bottom of flap. Photo taken on 25 May 1970 by Fred Roos. NF-R133-8.

PHOTO BELOW: T-28B, 137780, from the Test Pilots School with NATC on tail, and badge on cowling. Photo by A. Hafter via R. Besecker, June 1965.

MCAS
NEW RIVER
0047
MARINES
140047

03
BEAUFORT
8134
MARINES
MARINES
138134

QUANTICO
8173
MARINES

Authors Note: In the late 60's-early 70's, every AOC-AVROC, upon graduation, expected to be issued his own brand new shiny Phantom, and orders to Nam. In that era when every young Ensign drove either a Vette or comparable fast car in the 400 h.p. power class, the T-28 stole their hearts. When the T-28 entered the landing pattern, every one would watch. Coming in hot and heavy (shades of John Wayne) with power on, then poping the speed brake and pulling into a steep bank to bleed off the airspeed and enter the pattern. It was much more exciting than watching a Phantom come in like a lead brick. At idle a T-28 was music to the ears of all those hot-rodding Ensigns. The popity-popity sound of 1425 h.p. at idle was much more impressive than the whine of the F-4s jet engine.

PHOTOS TOP TO BOTTOM:
T-28B, 138130, from MCAS New River. Photo via Berger.

T-28B, 138341, station aircraft NAS Meridian, in 1974. Photo via Berger.

T-28B, 137641, from reserve base NAS Grosse Ile, Mich. Photo by R. Harrison.

T-28B, 137734, from reserve base NAS Anacostia. Photo taken on 11 May 1962 by R. Besecker.

PHOTOS TOP TO BOTTOM:
T-28B, 137640, MCAS Quantico taxying out. Note how pilots sit in open cockpit. Photo taken on 4 May 1961 by Frank C. Hartman via AAHS.

T-28B, 137708, NAS Memphis. Note silver prop hub, 708 under wing, natural metal leading edges and Memphis under Navy. Photo by J. Wible via R. Besecker.

T-28B, 137682, NAS Memphis. Photo by Nick Williams via R. Besecker.

T-28B, 137668, from Fleet Marine Force Atlantic. Note day-glo check mark on tail with black outline. Photo from Lionel Paul collection.

T-28B, # unknown, of H&MS-10 in overall yellow with dark green fuselage and wing stripe. Silver prop and hub with chrome yellow tips. Grey anti-glare panel and black exhaust stain area.

PHOTOS AT LEFT, TOP TO BOTTOM: T-28B, 138289, station aircraft from NAS Barbers Point. Note words Barbers Point under wing. Photo by R. Harrison.

QT-28B, 137663, drone director aircraft based at NAMC Point Mugu. Fuselage color is dark blue with day-glo red tail and rudder. Bright yellow wings and horizontal stabilizers. Wing has day-glo red strip outboard of main landing gear. Note black silhouette of SR—71 on tail.

Earlier photo of 137663 with T-28BD designation on fuselage. Colors the same as above except wing stripe is missing and the white around the number 7663 on tail. Both photo from Lionel Paul collection.

VA-122

PHOTOS TOP TO BOTTOM:
T-28B, 138340, Taxying at NAS Miramar during May 1970. Photo from Lionel Paul collection.

T-28B, 137651, on March 1966. Overall white aircraft with natural metal leading edges. Photo by N. Taylor via R. Besecker.

T-28B, 138271, in grey and white scheme, with yellow fin stripe and black outline.

T-28B, 138245, on 14 August 1970. Photo by Fred Roos NF-R192-12.

VT-2

VT-2 was commissioned on 1 May 1960 at NAS Whiting Field and operated the T-28B aircraft until it was replaced in the late 1970s by the T-34C.

PHOTO AT RIGHT: T-28B, 138199, of VT-2 on 11 Jan. 1963. Photo via AAHS.

PHOTO BELOW: T-28B, 138117, on 15 March 1970. Note red nose wheel and position of 2G under wing. Photo by Fred Roos NF-R83-6.

PHOTO BELOW: T-28B, 138259, in the original all yellow paint scheme. Note the fragmentation of the number and letter codes as well as slanted numbers 593 on tail and small number 3 on cowling. Photo was taken in July 1958 and although the colors have not yet been changed to red and white, the tail code has been changed to 2W for VT-3. Photo by A. Gelbudas via R. Besecker.

VT-3

VT-3 like VT-2 is stationed at NAS Whiting Field. The two squadrons mission and flight syllabus for training Naval Aviators in the T-28 were the same. This mission consisted of the basic training of student pilots in five phases of flight instruction: transition, precision/acrobatics, basic instruments, radio instruments, and formation tactics.

White circle with red shield. VT- 3 lettering white on black.

PHOTOS TOP TO BOTTOM:

T-28B, 137716, with 134 on tail surrounded by white, natural metal leading edges and VT-3 emblem on cowling. Photo by R. Harrison.

T-28B, 138155, of VT-3 in the mid 1970s with red rudder added and white codes replacing the black ones. Photo from Lionel Paul collection.

T-28B, 137766, in 1974 with Comtrawing Five written on tail in red on a large white check mark. Note speed brake. Photo via Berger collection.

T-28B, 138149, of VT-3 with emblem on cowling. Photo by Interair.

VT-6

VT-6 has the mission of training pre-helicopter students in the art of instrument flying. In this capacity VT-6 flys the last of the Navy's T-28s (except VT-27) which were taken out of storage at MASDC to continue the squadrons mission.

Historically VT—6 provides the student with a four week course in radio instruments and cross-country navigation. Instruction consists of: 35 hours of ground school, 8 hours of link training, 6 hours of flight procedures and 11 flights totaling 20 hours. Over the years VT-6 has maintained an average of 20 aircraft on hand.

PHOTOS BELOW: top — T-28B, 140022, VT—6 on 11 Oct. 1973. Note open oil filler access and position of numbers 222 under the wing. Photo by Fred Roos, NF-R618-6. Middle — T-28B, 140022, again in 1980 with the new Training Air Wing Five code letter E on tail, and decked out in grey and white color scheme. Photo from Lionel Paul collection. Bottom — T-28B, 138149, of VT—6, TAW-5, stationed at NAS Whiting Field in Aug. 1980. Photo by Fred Roos, NF—R894-2.

PHOTO ABOVE: T-28B, 140016, TW—5, April 1978. Photo by R. Harrison.

PHOTO ABOVE: T-28B, 137749, of VT—10. VT—10 provides basic training for student Naval Flight Officers and Air Intelligence Officers and has flown a mixed bag of aircraft. Photo by R. Besecker.

T-28C, # unknown, from Basic Training Unit Three, NAS Saufley Field. Overall yellow with silver prop and yellow tips.

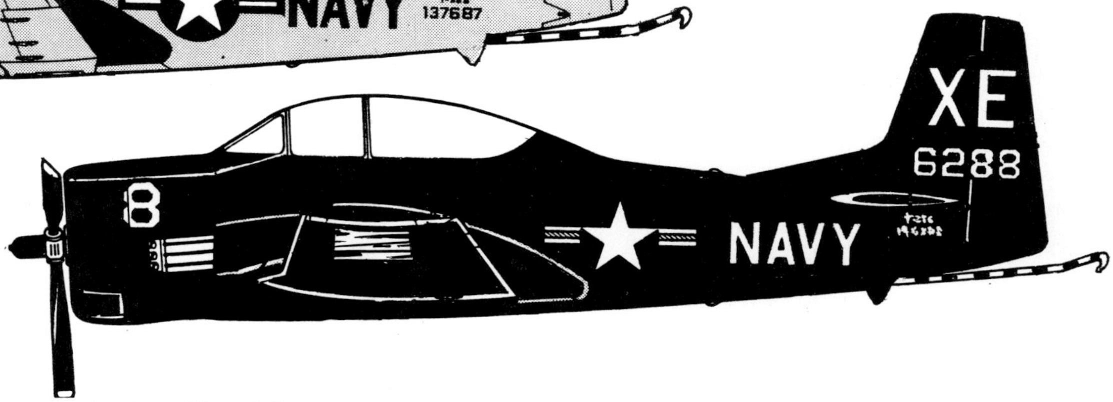

T-28C, 146288, of VX-5 in overall Navy blue scheme.

PHOTO BELOW: The first T-28C, 140053, landing. Note fully extended speed brake and landing gear. National Archives Photo.

T-28C

The T-28C, which made its first flight on 19 Sept. 1955, grew out of the Navy's T-28B as a carrier capable aircraft. The C differed from the B in having a notched tail with tailhook, strengthened landing gear, and a shorter prop. In all 299 T-28C aircraft were built for the U.S. Navy.

T-28C Aircraft

Navair Numbers	Manufacture Sequence
140053 - 140077	1 - 25
140449 - 140666	26 - 243
146238 - 146293	244 - 299

T-28C GENERAL ARRANGEMENT

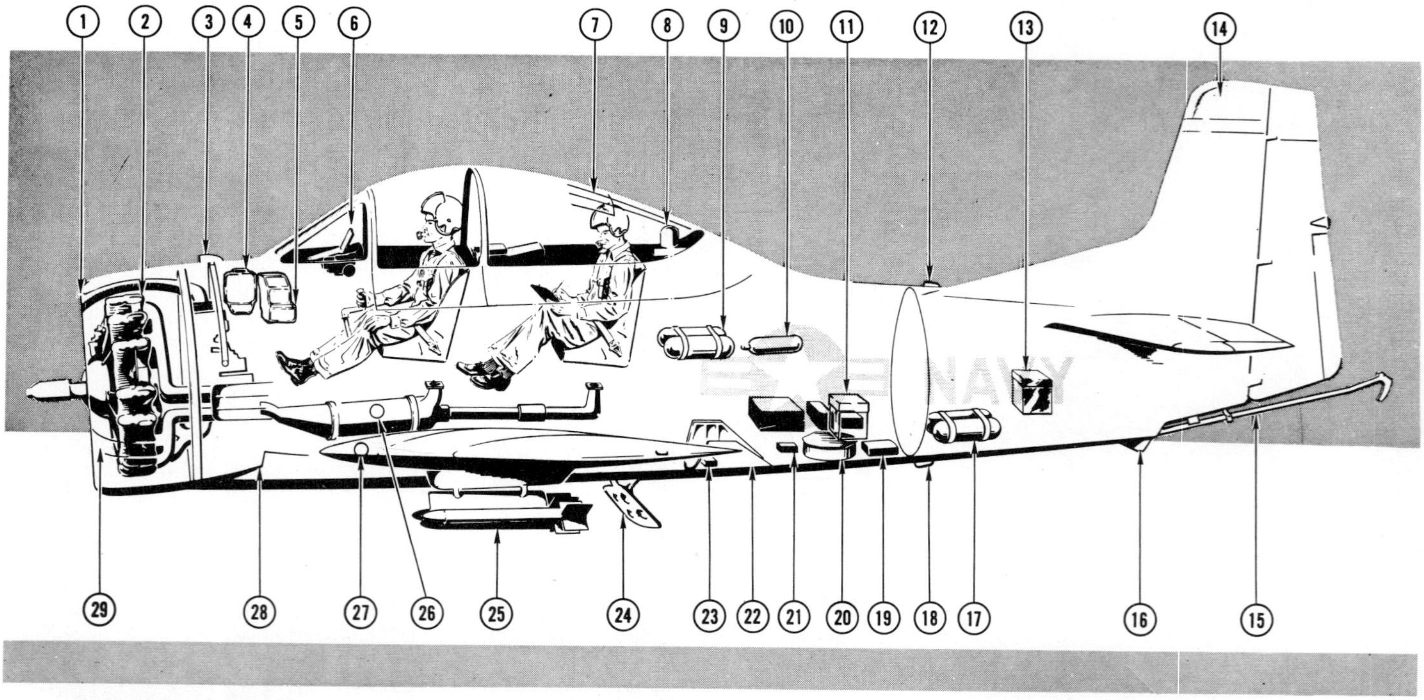

1. CARBURETOR AIR INTAKE
2. R-1820-86 OR -86A ENGINE
3. MAGNETO AIR INTAKE
4. ENGINE OIL TANK
5. HYDRAULIC RESERVOIR
6. MK-8 GUNSIGHT
7. AN/ARN-6 RADIO COMPASS SENSE ANTENNA
8. AN/ARN-6 RADIO COMPASS LOOP ANTENNA
9. OXYGEN CYLINDERS (T-28C)
10. CANOPY EMERGENCY AIR BOTTLE
11. BATTERY (T-28C)
12. UPPER ANTI-COLLISION LIGHT
13. BATTERY (T-28B)
14. UHF COMM/TACAN ANTENNA
15. ARRESTING HOOK (T-28C)
16. TAIL BUMPER
17. OXYGEN CYLINDERS (T-28B)
18. LOWER ANTI-COLLISION LIGHT
19. MARKER BEACON ANTENNA
20. AN/ARA-25 ADF ANTENNA
21. EXTERNAL POWER RECEPTACLE
22. BAGGAGE COMPARTMENT DOOR
23. DATA CASE (BAGGAGE COMPARTMENT)
24. SPEED BRAKE
25. ARMAMENT PACKAGE (2.25-INCH ROCKETS SHOWN)
26. COCKPIT HEATER COMBUSTION AIR OUTLET
27. COCKPIT HEATER COMBUSTION AIR INLET
28. OIL COOLER OUTLET
29. OIL COOLER INTAKE

ARMAMENT SYSTEMS

ARMAMENT TRAINING PACKAGE, AERO 1A

For training purposes, the aircraft may be fitted with an armament training system, Aero 1A, incorporating a Mark 6, Mod 0 fire control system and utilizing various armament packages. The armament control panel (figure 8-4) mounts in the center of the floor directly beneath the instrument panelin the front cockpit. The armament package mounts externally under each wing inner panel (figure 8-1) and consists of two gun packages or two bomb racks. Each rack will accommodate one bomb, one Aero 4B practice container, or a launcher with three rockets. A gun camera is mounted in the leading edge of the left wing, when guns or rockets are installed. Mixed combinations of armament cannot be carried. Strakes are required on each side of the fuselage of T-28C aircraft while carrying external stores.

GUN PACKAGES

A detachable gun pod containing one Type M2 or M3, .50-caliber machine gun, 100 rounds of ammunition, and a pneumatic gun-charging system can be installed at each inboard wing station. The charging systems are controllable from the front cockpit, and each has sufficient capacity to charge its respective gun about 20 times. When the guns are fired, the charging systems are inoperative, preventing accidental simultaneous gun firing and charging. The gun pods cannot be jettisoned from the aircraft.

BOMBS AND ROCKETS

To deliver bombs or rockets, the aircraft is configured with a modified Mark 51 Mod 14 bomb rack at each of the two external stations. Each rack is capable of carrying an Aero 4B practice container, with either eight MK 23, eight MK 43, or six MK

5 practice bombs, one AN/M38A2 practice bomb, one AN/M2A2 practice cluster, or equipped with rocket launcher, three 2.25-inch SCAR rockets. Bomb and rocket selection, arming, release, and firing are accomplished electrically by controls in the front cockpit. Sighting is done with the Mark 8 Mod 5 sight and individual buttons on the stick grip release bombs or fire rockets.

CAMERA

A Type AN-N6 or -N6A camera can be installed in the leading edge of the left wing to photograph the target. The camera, powered by the armament bus, operates automatically when the gun trigger or rocket-firing button is depressed and, in addition, continues to photograph the target for a period of 1 to 5 seconds after the trigger or rocket-firing button is released. The time of operation after trigger or rocket-firing button is released can be varied by the ground crew. No camera adjustments can be made by the pilot during flight.

TOW TARGETS

Tow targets of the sleeve type or the banner type may be utilized by installation of the tow-target equipment. Refer to Tow-Target Equipment, in this section.

External Stores

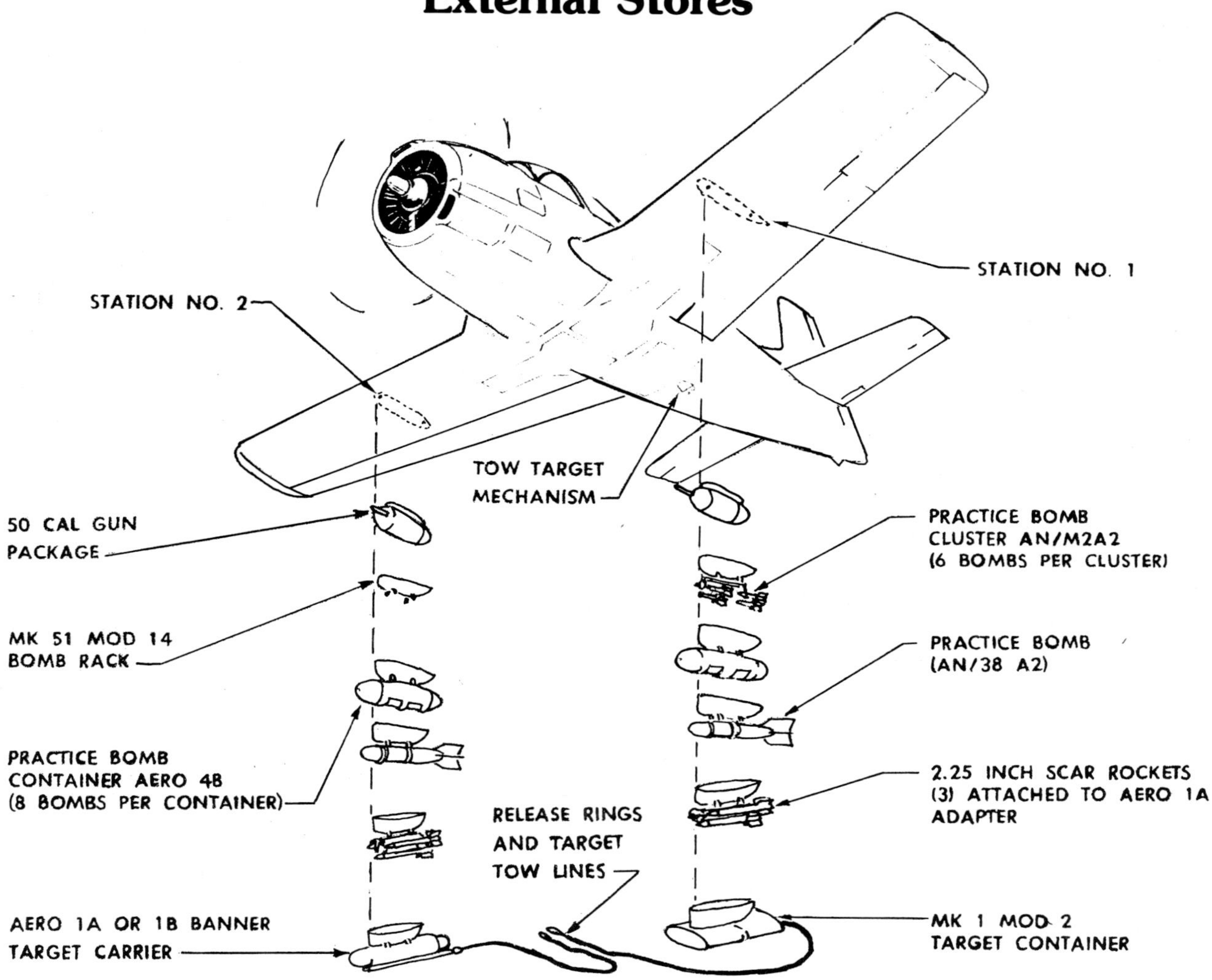

ARMAMENT (MODIFIED T-28C)

Some T-28C aircraft have been modified to accommodate a wide range of stores selection. Modifications include six store stations, three on the lower surface of each wing to accommodate various complements composed of fragmentation, general-purpose, napalm, and practice bombs; 2.75- and 5-inch rockets, and .50-caliber gun pods. A modified armament control panel (figure 8-5), utilized with the Mark 6 Mod 0 fire control system, supplies control of the armament system on the modified aircraft. Strakes must be installed when carrying external stores. See figure 8-2 for external stores loading capability.

WING STATIONS

The six wing stations mount armament packages that may consist of two M2 or M3 .50-caliber machine guns and pneumatic gun charger packages and four Aero 15C/15C-1 bomb rack/rocket launchers with adapters, or the gun packages may be removed and replaced by Aero 15C/15C-1 bomb rack/rocket launchers with adapters. Each station may be used for numerous stores within the station weight limitations, and mixed combinations of stores may be carried symmetrically.

BOMBING SYSTEM

The Aero 15C/15C-1 removable bomb rack has been adapted to the six store stations. The racks are capable of carrying stores of recommended type within the weight limitations of each store station. See figure 8-2. Six bomb select switches, one for each store station, are located on the aft side of the wheel wells. When the Aero 4B practice bomb container is carried, one of the wheel well bomb select switches must be placed in the disable position, for each practice bomb container carried, to preclude inadvertent dropping of the container. The practice bomb containers may still be emergency jettisoned. Bomb selection, arming, and release are accomplished electrically by the controls in the front cockpit. Bomb sighting is done with the Mark 8 Mod 5 sight and bombs are released by the bomb release button on the control stick grip. All other bombing system controls are on the armament control panel.

PHOTO BELOW: T-28C #1, 140053, takes to the air on its first flight, 19 Sept 1955. Note factory fresh yellow paint scheme and the extended speed brake. Photo by North American.

ROCKET SYSTEM

Each of the six wing store stations can accommodate various rocket pods containing several 2.75-inch FFA rockets. The rockets and pods are mounted on an Aero 15C/15C-1 combination bomb rack and rocket launcher. Rocket selection and firing are accomplished electronically by controls in the front cockpit. Rocket stations are fired individually, in pairs, or in salvo, depending upon the setting of the station selector switch. Aiming is accomplished with the Mark 8 Mod 5 sight and the contents of a single pod or pairs of pods are ripple-fired with each depression of the rocket-firing button on the stick grip. The gun camera operates automatically during rocket firing.

FIRE CONTROL SYSTEM

The Mark 6 Mod 0 is a pilot-operated fire control system providing lead angle information for precision aiming of the fixed guns, bombs, or rockets.

PHOTO AT LEFT: T-28C, 140548, MCAS Quantico, on 23 April 1974. Note battery access door and oil access door are open. Also note the words Quantico and Marines are slanted on the fuselage. Photo by J. Sullivan via R. Besecker.

Modified T-28C External Stores

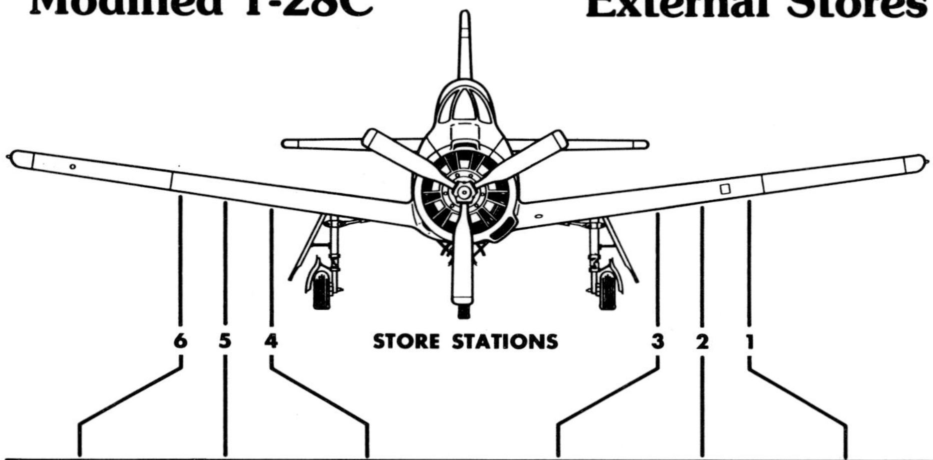

POUNDS MAXIMUM LOADING	150	250	500	500	250	150
STORES						
.50 CALIBER GUN PODS						
AN-M1A2 BOMB CLUSTER WITH 6 20-POUND FRAGMENTATION BOMBS						
AN-M28A2 BOMB CLUSTER WITH 24 4-POUND (MK 85) FRAGMENTATION BOMBS						
AN-M30A1 GENERAL PURPOSE BOMB						
AERO 4B CONTAINER WITH 8 MK 42 PRACTICE BOMBS						
AERO 6A1 LAUNCHER WITH 7 2.75 FOLDING FIN AIRCRAFT ROCKETS						
5-INCH HIGH VELOCITY AIRCRAFT ROCKETS						
IMPROVISED NAPALM BOMB USING 100-POUND PRACTICE BOMB BODY WITH IGNITER AND FUZE						
STATION FIRING SEQUENCE—SINGLES	4	2	6	3	5	1
STATION FIRING SEQUENCE—PAIRS	1	2	3	3	2	1

THE ABOVE STORES MAY BE SYMMETRICALLY MIXED WITHIN STORE STATION WEIGHT LIMITATIONS.

T-28C-1-63-1A

NAVY

NAVY
138185

PHOTOS: top — T-28C, 140514, of VA-122 in grey and white scheme, with yellow fin flash. Photo by Paul Stevens via Menard. Middle — T-28C, 140553, of VT-3, with white codes over the red. Note cowl flaps full open. Photo by Fred Roos, NF-R618-12. Bottom —T-28C, 140573, of VT—3, with faded paint and squadron emblem on cowl. Photo by N. Taylor via R. Besecker, 24 March 1966.

PHOTOS AT LEFT TOP TO BOTTOM: All yellow T-28C of NATC Tactical Test division leaves a carrier deck. Note 187 and TT on nose and cross painted on nose wheel. North American photo.

T-28C, 140657, NAS Quonset Point, June 1971, via Lionel Paul collection.

T-28C, 140063, NAS Pensacola, 9 April 1978. Photo by R. Harrison.

T-28C, 140061, NAS North Island. Note rocket tube mounted under wing. Photo by Fred Roos, NF-R528-4, 14 May 1973.

PHOTOS AT RIGHT TOP TO BOTTOM: top — T-28C, 140057, of VT-5 in flight. USN photo.

MIDDLE — T-28C, 140549, VT-5, USS Lexington, in very faded white and red scheme. Photo taken in March 1966 by N. Taylor via R. Besecker.

BOTTOM — T-28C, 140631, VT—5, USS Lexington, in last paint scheme. Color photo of this scheme on the back cover. Photo by Fred Roos, NF-R339-11, on 31 July 1971.

VT-5

Although now extinct VT-5, based at NAS Saufley Field, was the main user of the T-28C. VT-5s mission was to train students to hit the boat and become carrier qualified Naval Aviators. Qualification required the student to make two touch-and-go and six arrested landings.

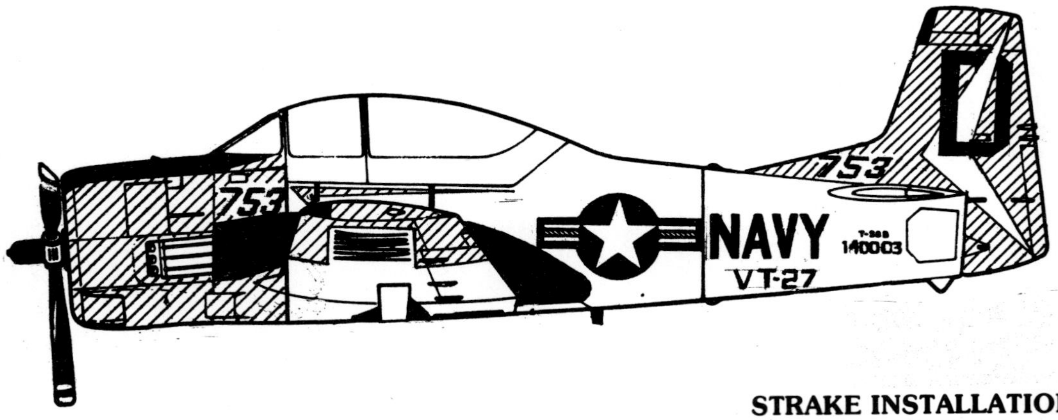

T-28B, 140003, of VT-27 on 22 Dec. 1979. Trainer white and red with a white check mark on the tail and a red check mark under the canopy.

STRAKE INSTALLATION

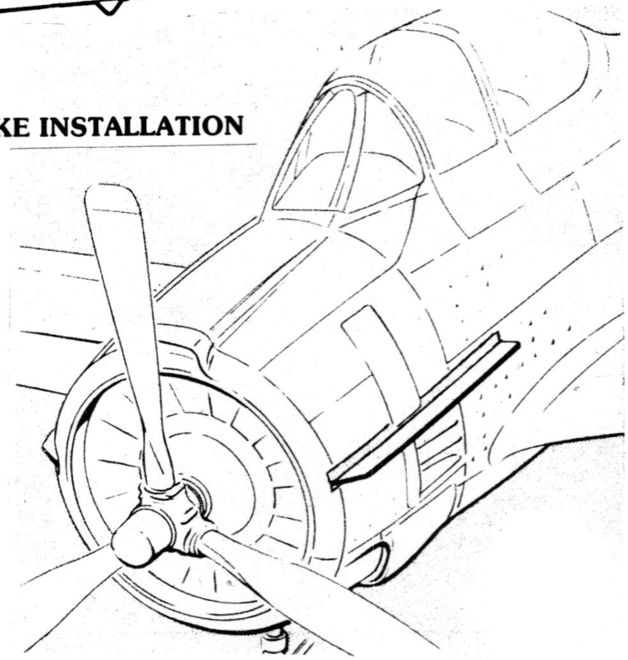

The three photos below taken in Sept. 1980 at NAS Pax River show civilian T-28C in grey and white with authentic markings. The three lines under the horizontal stabilizer read: NX289RD, then T-28C, then 140576. Plane also has UT-5s logo painted under left side of canopy. Photos by Jim Burridge.

Note in the photos and drawing above, the strake installed across the cowling. This indicates that this T-28C was one of the few modified T-28Cs. The NATOPS manual says "when gun packages or other external stores are installed, strakes must be attached. With the packages and strakes installed, spin entry and spin recovery are normal."

PHOTOS: top — T-28C, 146267, station aircraft NAS Pensacola. Standard red and white scheme with yellow rescue arrow. Middle — T-28C, 140601, station aircraft NS Saufley, note Indian head logo on nose. Face is darker red than background, with white headdress and black feathers. Logo probably derived from the fact that Saufley field used two radio call signs, Sue falls or Sioux falls. Note also cowling lip is black. Bottom — T-28C, 146239, of VT-3 with the new white lettering and E tail code over the red markings. Photo by Interail.

ARMY T-28B

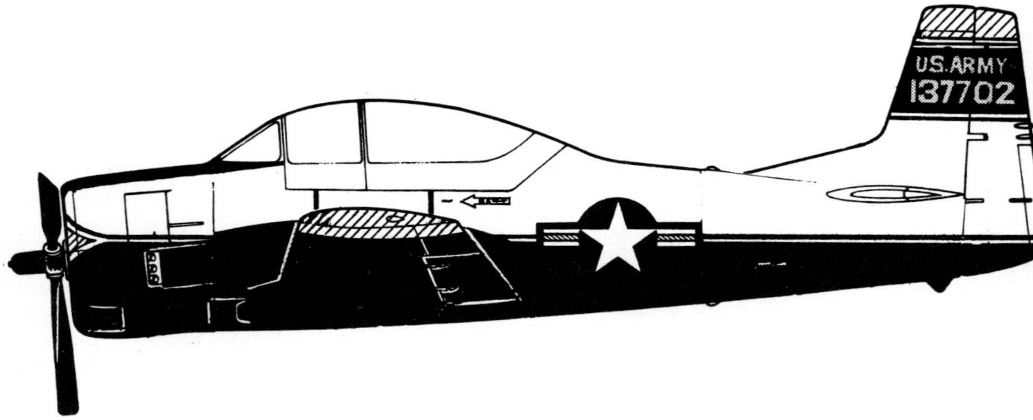

Drawing of U.S. Army T-28B — Fuselage upper surfaces white, lower surfaces and wing and tail band, extra dark gloss olive. Wing tip and fin tip red. Lettering U.S. Army and 137702 are yellow.

This page depicts U.S. Army T-28Bs used in the chase plane role for helicopter test programs.

PHOTOS: Top — T-28B, 138336, overall white with dayglo fin, rudder, outer wing panels and cowling. Photo by R. Esposito via R. Besecker, Jan. 1965. Middle — T-28B 138327, in red and white scheme, Edwards AFB, Nov. 1975, Author. Bottom — A different T-28B in the same scheme in Nov. 1976, Author.

CIVILIAN BLUE ANGEL

The three photos below, taken in Sept. 1980 at NAS Pax River show T-28B, N100JE, painted with full Blue angel markings. Plane is overall Blue Angel blue with chrome yellow lettering and wing, tail and fin tips. The lip of the cowling, the outer halves of the prop blades, and the panels behind the exhaust stacks are highly polished metal. The inner halves of the prop blades are grey and the prop tips are yellow. The gear is painted white. The name on the left front canopy rail is J M ELLIS III. Note non-standard spinner. Photos by Jim Burridge.

TWO TOP PHOTOS — Two views of a re-engined T-28A, painted up to look like a factory fresh T-28B. Aircraft is yellow overall with white gear and silver prop with yellow tips. T-28 is registered N2800W with NAA logo on tail and a shaggy dog painted on gear door. Name under canopy is Col. Tom M. Neal with experimental below that. In the top photo you can see the T-28A roll-over truss. Photos taken at Chino, Calif. by Don Fetters, May 1981. MIDDLE PHOTO — T-28A, N7491C, rotting away at Chino in the late 70s. Fuselage top white, bottom blue grey. BOTTOM PHOTO — T-28A, N9611C, of the CAP at Van Nuys, Calif. in the late 70s. Plane is painted in Air Force two-tone green and tan. Rudder trim tab has red-white-blue stripe and "N" number is yellow surrounded by white. Photos by Author.

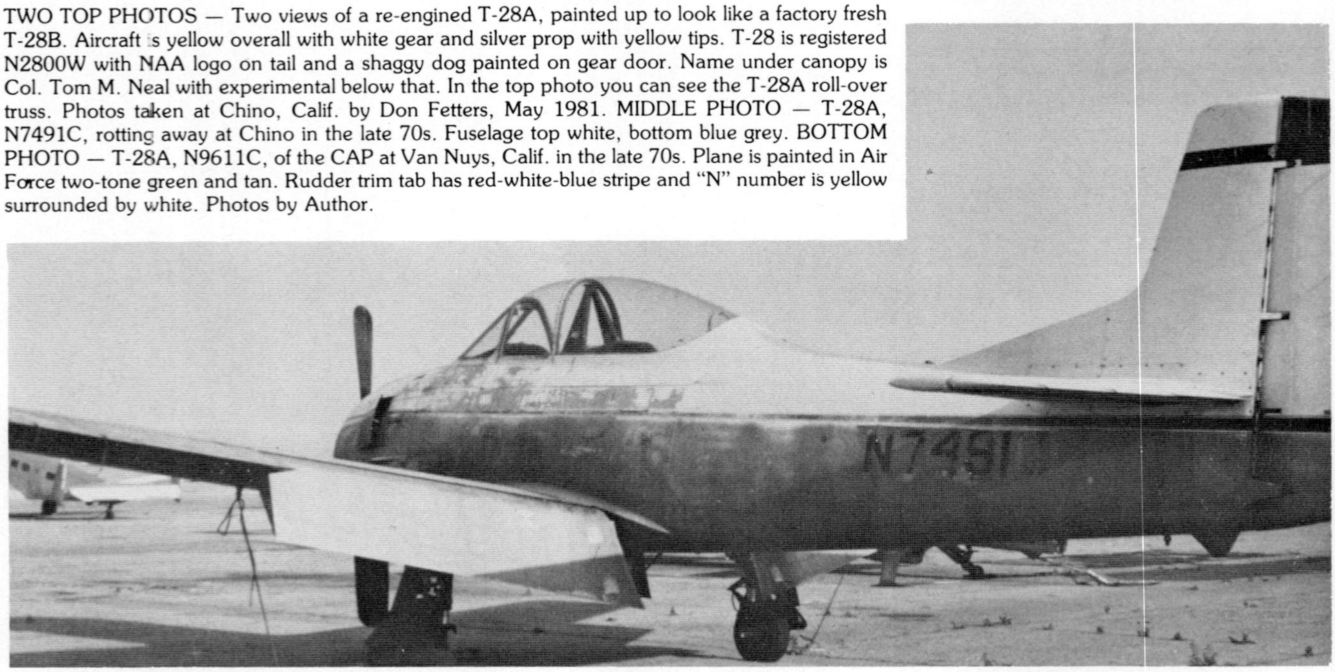